Global Warming and Climate Change

Edited by John P. Tiefenbacher

Published in London, United Kingdom

IntechOpen

Supporting open minds since 2005

Global Warming and Climate Change
http://dx.doi.org/10.5772/intechopen.84934
Edited by John P. Tiefenbacher

Contributors
Isabel Martínez, Gemma Grasa, Jose R. Fernández, Angela Oyilieze Akanwa, Ngozi Joe-Ikechebelu, Nelson Chanza, Veronica Gundu-Jakarasi, Nizar Jaoua, Paritosh C. Kulkarni, John P. Tiefenbacher

First published in London, United Kingdom, 2020 by IntechOpen
IntechOpen is the global imprint of INTECHOPEN LIMITED, registered in England and Wales, registration number: 11086078, 7th floor, 10 Lower Thames Street, London, EC3R 6AF, United Kingdom
Printed in Croatia

British Library Cataloguing-in-Publication Data
A catalogue record for this book is available from the British Library

Additional hard and PDF copies can be obtained from orders@intechopen.com

Global Warming and Climate Change
Edited by John P. Tiefenbacher
p. cm.
Print ISBN 978-1-83880-994-2
Online ISBN 978-1-83880-995-9
eBook (PDF) ISBN 978-1-83880-996-6

For EU product safety concerns:
IN TECH d.o.o., Prolaz Marije Krucifikse Kozulić 3, 51000 Rijeka, Croatia, info@intechopen.com or visit our website at intechopen.com.

Meet the editor

Dr. John P. Tiefenbacher (PhD, Rutgers, 1992) is Professor of Geography at Texas State University. His research has focused on various aspects of hazards and environmental management. Dr. Tiefenbacher has published on a diverse array of topics that examine perception and behaviors with regard to the application of pesticides, releases of toxic chemicals, environments of the US–Mexico borderlands, wildlife hazards, and the geography of wine. More recently his work pertains to adaptation to climate change, spatial responses of wine growing to climate change, the geographies of viticulture and wine, artificial intelligence and machine learning to predict patterns of natural processes, historical ethnic enclaves in American cities and regions, and environmental adaptation by European immigrants to North America.

Contents

Preface

2019 was the second warmest year on record. The US National Aeronautics and Space Administration reported in 2019 that, based on mean annual global temperatures, the most recent five years (2014–2018) had been the warmest on record since 1880. Each of those years was approximately 0.75–0.80°C above the baseline mean over that stretch of time. Including 2019, 18 of the last 19 years (i.e., since 2000) have been the hottest on record as well. Despite the commitments made by the 195 states that signed the Paris Agreement in 2016, the concentration of carbon dioxide in Earth's atmosphere has continued to rise. As of March 2020, atmospheric CO_2 has reached a concentration of 414 ppm. The people of Earth are heading at an increasing rate toward a crisis as 1.5°C warming by 2030 is looking very certain. The Kyoto Treaty was based on hopes of preventing 2.0°C by the end of the century, but at the rate of current progress, that will likely happen by 2050. With the withdrawal of the United States from the Paris Agreement in November 2019, and the weak action to meet the commitments of many of the remaining signatories (with a few exceptions), the future, at present, looks bleak.

Global warming is gaining attention in the countries of the world that had previously viewed the issue as someone else's problem. Americans used to consider themselves immune to the matter, seeing it as a "Third World" problem. That attitude is shifting dramatically as the public has come to see the threats that global warming poses. While still maintaining a poor understanding of the connections between global atmospheric warming and the natural systems (hydrological, ecological, atmospheric, etc.) they interact with, there is an understanding that the future will be filled with a growing number of challenges to meet their basic needs, to stay healthy, and to remain comfortable.

Two European countries (the United Kingdom and Germany) have made great progress toward mitigating their national carbon footprints, however. In 2019, the United Kingdom announced that they had reduced their greenhouse gas output by 38% below their 1990 emission levels (44% in 2018, however). They were approaching the halfway point to reach their 2050 goal to reduce emissions by 100% of their 1990 production. Germany is also aiming for carbon neutrality. In 2018, they had reached a reduction of about 30% from their 1990 level and they are striving to reach 40% by the end of this year.

The United States, on the other hand, were approximately equivalent to the greenhouse gas output of 1990 in 2016 but since the beginning of the Donald J. Trump administration, the level has risen by 4–5% to about 5.5 $GtCO_2eq$. Currently, the trajectory is roughly flat, which in no way helps the people of the United States or the rest of the world. China's emissions are up about 350% from 1990 at about 14 $GtCO_2eq$ and the trajectory is rising. Of course, the footprint of China is greater than the United States, but per capita, they are roughly equivalent. Together, these two states contribute more than 40% of the global greenhouse gas load. Lack of improvement from these two powers will keep the rest of the world in "hot water."

In this context, this book seeks to highlight aspects of the problems associated with global warming and the spin-off consequences that it generates, principally matters associated with changing climates. It seeks to explore topics related to consequences, responses, and efforts at changing the trajectories of the challenges. Indeed, there are six chapters that discuss various aspects of global warming and climate change. They are organized into two sets. The first set of three chapters pertain to thinking about these problems. The second set of chapters regards actions taken to improve processes that mitigate the impacts of technologies and community activities.

Managing Minds

In the introductory chapter, I discuss an issue that seems to impede public support for calls to action in the United States, and perhaps in other countries of the world as well. That people, some of whom are public educators (not necessarily teachers, but people like broadcasters, journalists, politicians), even educated and seemingly intelligent people, don't seem to understand that they do not fully understand the matters at hand, particularly in that they unknowingly portray the concepts of global warming and climate change inaccurately. In trying to clarify these terms, I try to present the implications of understanding and not understanding.

In Chapter 2, Angela Akanwa examines actions that foster environmentally conscious living, reflecting the fact that global warming and climate change are merely symptoms of a deeper socioeconomic and cultural affliction associated with development. She examines the approaches that could be adopted and applied in Nigeria to meet goals that not only contribute to remediating global warming and climate change, but also "green" Nigerian society and its landscapes to approach sustainability.

And in Chapter 3 by Nelson Chanza and Veronica Gundu-Jakarasi, we explore the consequences of responding to global warming mitigation and adaptation to climate change in Zimbabwe. Chanza and Gundu-Jakarasi critically examine the claims of benefits to various economic activities (agriculture, water production, energy, and health) achieved by government policies, responses, and interventions in these sectors. They argue that though there are some advances made by government programs, there are many unfulfilled promises, several areas of opportunity for new development, and the prospects for successes. They question whether the mitigation, adaptation, and development efforts are sustainable in the long run.

Managing Activities

In the fourth chapter, Nizar Jaoua discusses and demonstrates modeling of global emission reductions to meet the United Nations' carbon dioxide concentration goals. Planning smooth pathways to achieve reduction targets that slow global temperatures and hold them to near the 1.5°C maximum warming, while still enabling economic systems to transition to alternative, low- to no-carbon energy systems, is of vital importance to avoid catastrophic changes.

The fifth chapter by Paritosh Kulkarni presents the results of a study of a process to reduce the production of black carbon or soot from diesel engines by introducing chemical impurities into the fuel. Black carbon is the second greatest contributor to global warming. It accounts for roughly two-thirds of the warming caused by carbon dioxide and twice as much as methane. Though not strictly a greenhouse gas, atmospheric soot contributes by absorption and retention of direct incoming solar radiation, rather than capturing reradiated energy from Earth's surface as greenhouse gases do. Thus, reduction of soot emissions can have a significant impact on global warming rates.

Chapter 6, by Isabel Martínez, Jose R. Fernández, and Gemma Grasa, discusses a technological process to enable the production of hydrogen fuel. Hydrogen fuel is notoriously difficult to produce due to high energy costs and production of carbon dioxide, which counteracts the advantages that hydrogen fuel technologies promise. Aiming for carbon-free energy technology, the carbon-capture process using calcium and copper reactions sequesters the carbon dioxide from the waste stream. The promising advances of the new technology would soften the transition to carbon-free transportation.

It is hoped that this collection offers new insights and new ideas that may promote interest, action, and further research on topics related to global warming and climate change.

John P. Tiefenbacher
Texas State University,
USA

Section 1

Managing Minds

Chapter 1

Introductory Chapter: Climates, Change, and Climate Change

John P. Tiefenbacher

"There is an infinite amount of hope in the universe ... but not for us."
Franz Kafka, c. 1920

1. Introduction

Global warming is no hoax. It has been amply substantiated [1]. That is not to say that "science" knows everything there is to know about global warming, only that there is no doubt that it is happening and that it is indisputably due to human activities that have loaded unnatural levels of greenhouse gases into Earth's atmosphere over the last 200 years [2]. Global warming is generating many significant challenges that will affect humans' superficial comforts and threaten the foundations of our survival [3, 4]. Changing climates are only one of the complications that we will face. Some of the other complications are: rising sea levels; acidifying oceans; diminishing extents of components of our cryosphere, particularly glaciers, permafrost, Greenland's ice sheet, and the ice cap of Antarctica; changing distributions of fresh and saltwater; changes in habitat size (shrinking for native species and growing for invasive species) and distribution; the spreading of diseases that have been limited by climate conditions of the past; destabilization of ecological systems, particularly the loss of coral reefs; mismatches between soils and climates, hydrological patterns, plant and animal life, weather processes, and seasonality undermining global and local food production; and changing patterns of hazard related to and linked to all of these impacts that will dislocate and force relocation of human populations, causing further tumult [5].

While scientists have examined many analogs to the prospective consequences of global warming by studying isolated processes on isolated places at times when they were of rather limited concern, many of the emerging changes are challenging the limits of knowledge and understanding of how Earth's natural systems function. We often lack detail that might allow us to "predict" (we really need to be able to *project* the expected changes onto today's conditions and into) the future precisely and accurately so that we can design, plan, and direct our collective life-trajectories toward survival practices that are sustainable. The task is clear enough to know that most human beings have myopic, narrow, limited understandings and views of the consequences of global warming, and an even narrower and substantially superficial view of what climate change means to their lives, and what it means for the future for humans [6, 7].

This chapter discusses the concepts behind understandings of global warming and climate change. It emphasizes the need to encourage change that not only mitigates the behaviors that are contributing to the problem of global warming, but also promotes a deeper, more profound understanding of climate change, so that we

IntechOpen

can meaningfully probe the dark future to fathom what we can expect from global warming. The meaning of the term "climate change" may have already been lost as it has been commonly subsumed into the mistaken belief that *Earth's* climate is shifting to a new normal; like one is turning the dial and increasing the heat under the pot on a stove, inferring that it is simply a matter of turning down the flame. We must systematically obliterate and reconstitute its meaning in public discourse, so that an accurate meaning of "climate" and the ramifications of "change" can be applied to our worlds. As people come to viscerally understand climate change and its consequences, the change can be more intelligently imagined in terms of every geo-, bio-, social, and economic system one might depend upon, as well as on every product upon which one relies. The term "climate change" is used by some to scare (or at least motivate) people into "pro-climate" action [8] (because it is a threat to our existence), even though climate is not actually a tangible "thing" at all. It can be used and then casually dismissed by signifying that it is only a childish fear of a bogeyman (it is just a figment of *your* imagination) and that climate change is not real. Some arrogantly express their lack fear (because our might and our intellect make it easy to manage). The reality is probably far beyond either end of that spectrum: the changes we experience will be more profound than we can imagine and it would be easy to "fix" if we were to do what is needed and accept the long period of time for the world to right itself. But we cannot simply stop our greenhouse gas production and expect a miraculous return to normal (as many have long tended to believe) [9]. Normal is gone. And all of Earth's human and nonhuman inhabitants may not see a new "normal" for a very long time.

2. Climates, change, and changing climates

If one listens to or reads the media of journalists, commentators, and public servants (particularly politicians)—people from whom the public normally receives new information and upon which they (often) base their understanding of their lifeworld—it is evident that few of the messengers have accurate and clear grasps of the concept "climate," and yet they have a deep desire or feeling of responsibility to provide a clear explanation of climate to the public [10, 11]. The most egregious misapprehension of climate is that Earth has one (and only one). Our planet does not have "a climate." The climate is a conceptualization created to intellectually portray the combined conditions of temperature AND precipitation conditions of a region. Earth does have a "global" atmospheric temperature (this is how the globe's temperature can be said to be rising—global warming). But it is wrong to believe that there is a measure of global precipitation. In terms of water, Earth is a closed system. There is a fixed and finite amount of water on Earth and it circulates globally in all its forms (vapor, liquid water, and ice) constantly, we call this the hydrological cycle. Water changes state and spatial distribution continually because of seasons, atmospheric and oceanic circulation patterns, weather events, precipitation, evapotranspiration, and thermal conditions. The Earth has MANY climates. The number depends upon the mathematical detail, sophistication, and characteristics upon which climates are defined and distinguished.

The term "climate" is often (mis)used interchangeably with weather, particularly when people are talking about their personal, empirical (past and present) experiences of the conditions of the atmosphere in which they live. Their fundamental mistake is that they believe that it (climate) is a phenomenon, is tangible, that it is something "real" that people can viscerally experience, and that it can be sensed and measured in real-time. Climate, in fact, is not real. It is knowable through either of two methods: statistical analysis or inductive inference.

DOI: http://dx.doi.org/10.5772/intechopen.92192

2.1 Knowing climate statistically

Climates are usually defined statistically. Climate is a mental visualization, if you will, of atmospheric tendencies devised to explain the differences and similarities between (large and small) terrestrial (i.e., land) regions of the world. A climate-classification system allows one to categorize climates by averages and ranges of temperatures, available moisture, and weather phenomena over (*at least*) 30-year periods. The most meaningful climate-classification schemes are based on large datasets containing long records (again, at least three decades' worth) of weather data distributed over Earth's terrestrial surface (oceans do not have climates in our conceptualization). Daily thermal and precipitation records are used to characterize "normal" weather conditions (i.e., tendencies) at and near each weather station (which are proxies for larger zones in lieu of a dense array of instruments measuring the atmosphere). Climates are also characterized by means, extremes, and seasonal patterns distinguished as regular occurrences of major shifts, or of extreme conditions, like frosts or freezes, monsoons, and hydrological droughts that occur annually. Why were climates created? The most basic need was to discern the opportunities and challenges one might expect for day-to-day and long-term survival and comfort. Having and knowing climates establishes a basis upon which we can consider our life-prospects, particularly for planning future activities (getting water, growing or gathering food, keeping our bodies healthy and maintaining our comfort) in the context of weather and seasonal weather patterns.

2.2 Knowing climate through inference

Climate statistics, however, do not magically reveal the implications of weather data. Even long ago, when Greeks talked about torrid, temperate, and frigid zones, they were reflecting on the prospects for or challenges of life in other regions of the world (naming parts as "summer-less," "intermediate," and "winterless" might predict opportunities and limitations for agriculture). Modern climates are much more sophisticated and more complicated, as is our need to know whether our more sophisticated and more complicated activities can be safely or profitably conducted in places around the world. In the absence of weather data upon which a classification schema can be based, scientists and nonscientists before them inferred climate conditions based on the empirical evidence on the ground, particularly on the vegetation, the least mobile occupant of any environment. The vegetation that grows anywhere can logically be regarded as the plants that have survived the conditions in that place. By observing the compositions of plant communities and considering each plant's characteristics (anatomy, physiology, and hardiness), one might inductively determine (using higher order, more sophisticated understanding of plant biology) the thermal and hydrological conditions that have prevailed in that place. Major ecosystem types are often associated with (and they even supply the names for) the spectral product of these variables: rainforest, tropical savanna, desert, steppe, and tundra are terms that are often used to identify "climates." So, it might not be difficult to understand how someone might believe that because plants are used to name climates and because plants are evident in the landscape, climate must be apparent ... visible. We must be capable of perceiving climate right now.

The problem is, looks and logic can be deceiving. Some plants have features that may fit well in other places, in other ecosystems. Plants can be unnatural (due to invasion) in a place, perhaps promoted by natural and unnatural disturbances of landscapes. Some plants might have been introduced from other regions with markedly different climates. Transplants or invaders may be found outside of their normal zones, supported artificially for aesthetic purposes. Plants are not always the best indicators of climate. For example, certain characteristics (thick, moisture-rich tissue) of so-called succulent plants are commonly thought to be drought- and heat-resistant features. These plants might be most often found in arid and hot regions like deserts or in places that experience periods of drought each year. Every continent, other than Antarctica, is home to succulent species. But not all succulents reside in arid regions or in places having annual dry seasons. For example, *Opuntia humifusa* (the eastern prickly pear cactus) is found in southeastern Ontario, Canada (near Lake Huron) in the remnants of the Carolinian forest (a region that is certainly not a desert, certainly not hot, and not an arid place). Similarly, "evergreen" (non-deciduous) plants are found from the tropics to the subarctic (notwithstanding that some lack cones which distinguish conifers from other evergreens). In fact, humans have modified landscapes to the point where hydrophilic plants can be grown abundantly in desert climates, assuming that sufficient irrigation is provided. Plants (by themselves) are not perfect indicators of climate.

2.3 Recognizing climates is best left to science and scientists

Temperature and precipitation patterns and moisture and thermal regimes may, during any given week, month, year, or decade, depart from the norm and leave a false impression of a region's climate. Personal experience is not input into the process of climate classification. Only carefully and consistently collected data are used. One's personal observations about trends in weather (or climate) do not supersede (or even complement) scientific data because people are ill-equipped (eidetic memory or otherwise) to gather and analyze the factors upon which climates are based. Metrics of temperature and precipitation are significantly more precise, reliable, and consistent than personal observation. The data are also more durable. The data are likely to be different from place to place, and this creates patterns of difference across space. Some proximate places may have dramatically different, even contrasting, averages, extremes, and event frequencies. At some point at some distance from a weather station, the long-term conditions may be so different that they can reasonably said to be in different "climates." The variation of data gathered at set locations over time can be analyzed to tell us whether there has been change in local and regional climates.

Every location on Earth has a climate and locations are grouped into regions of similar conditions, ultimately yielding a global map of climate regions, which is a pastiche of similar and seemingly static conditions. It should be noted, however, that the limits of a climate are somewhat arbitrarily established (usually based on round numbers, like 20, 40, or 60 inches of precipitation, for instance). Periodically (perhaps each decade), the patterns of the local conditions can be reevaluated and the boundaries on the map of climate regions can be shifted to more accurately reflect the data for the most-recent 30-year period. This is usually done each decade (2020 is prompting a reconsideration of the map).

2.4 Knowing the agency of climate

Very importantly, "climate" is believed to determine many aspects of localities' natural environments. Climates (it is truer to say very long-term—centuries, millennia—weather conditions) influence soil development: soils form very slowly and reflect the prevailing physical and chemical conditions that affect weathering of the parent material and availability of organic matter from decomposing vegetation. Climate, then, also influences the types and abundance of flora and fauna that shape soil development. Climate dictates long-term water supplies. Plants, specifically trees, that live longer than the 30-year climate period also record the past's weather. They aid us in gaining an understanding of the distributions of past climates. Dendrological and palynological records provide additional pieces of information to the nature of the climates of the past.

Over very long periods (tens of thousands to millions of years), climates (and therefore soils) also reflect the planet's context—solar activity, axial rotation, revolution around the sun, and global events like asteroid impacts, eruptions of super volcanoes, periods of continental glaciation, etc.—which may destroy our ecosystems and perhaps even eliminate species from Earth. Climates of the past are imprinted on the landscape and in the lithology of the planet. These sources offer more, very lengthy records of the conditions of the distant past. It is the millennia-long records that enable us to compare past climates to contemporary climates to discern the radical modification of climates that is currently afoot. Climates are being wrenched from their old consistent conditions and the changes will have substantial influence on regional conditions into the distant future.

3. Change

As a verb, change can be either passive or active. It can refer to a one-time modification of one set condition to another setting. It is therefore passive, one and done (i.e., there was a change, or something has changed). Or it can refer to an ongoing process (in a sense, an evolution), a process that has not stopped and may never stop. Changing or being "in flux" can complicate circumstances when stable conditions (like a climate) are expected and are relied on to plan for short- or long-term futures.

3.1 Implications of change

For instance, to be successful in producing massive yields per hectare, modern commercial agricultural activities may require heavy commitments for capital and services that are investments based on forecasts of future production. To grow a specific crop may require specific tools, equipment, pesticides, fertilizers, minimum volumes of water, and other inputs. The fiscal nature of farming necessitates financial planning to survive each year and to survive in the long run. When the circumstances (both intrinsic—environmental, social, personal conditions, et al.—and extrinsic—conditions of markets, competition, labor and supply costs, consumer demand, et al.) of farming change, reevaluation of the plans is necessary. How these changes are occurring and what the long-term picture for the array of conditions may evolve are important aspects of the information that industrial agricultural producers desire for their decisions to continue, to modify their plans, or to quit the activity because they can expect it to fail, before they end up in debt. What if we can no longer be confident about our (old) assumption that "next year" will be much like this year, weather-wise? The extrinsic context of farming, in this example, is the foundation of the activity in which farmers participate. Most of these intrinsic and extrinsic conditions are anthropogenic and can be manipulated by people, by businesses, by trade, or by government actions and policies. But the environment is different—many environmental conditions, like seasonal or annual weather conditions, particularly extremes and changes to the periodicities and durations of conditions (growing seasons and plant phenology, for instance), may become insurmountable if they are continuing to change, particularly if the changes are unexpected or unpredictable [12]. Decisions may need to be made quickly, particularly if crop production is diminishing [13].

In the United States and Canada, as in many other parts of the world, the national and regional patterns of many economic sectors have evolved over several hundred years to fit into and thrive in specific environmental conditions, particularly regions' climates. More recently, agricultural development has used technological advancements and mechanical and economic efficiencies to reduce costs and increase profits and to establish roots in new places. We have seen the emergence of agricultural regions dedicated to specific crops (the Corn Belt, the Wheat Belt, the citrus-growing regions, fresh fruits and vegetables, wine growing) or to specific activities (dairying, cattle ranching) that coincide with climatic conditions that fit production best.

Modern societies and modern economies have been developed on assumptions about consistency in the patterns of nature, on the stability of natural systems, on the unwavering resilience of nature. But what happens when there are no longer patterns to depend upon?

3.2 Responding to change

It was observed during the development of the epistemological paradigm "systems analysis" in the 1960s that ecological communities can be stable in numerous alternative states [14]. Assuming the undisturbed landscape is the "ideal" state for a natural landscape, what happens when it is disturbed by humans' activities undertaken to live in or exploit that landscape? That disturbance of an ecosystem's conditions may lead to ecosystem responses, either returning to its "original" condition (i.e., a resilience response) or shifting to a new state with a new equilibrium of ecological relationships. The upshot of the revolution in thought was to clarify that there is extensive evidence of interactions between "systems" that produce positive and negative feedbacks. The experiences of pioneers, settlers, loggers, farmers, and landscape engineers demonstrate that people can have a profound influence on nature and that, though extinction of species or wholesale transformation of environments may not always result, human activities can cause significant disequilibria in the world [15].

Sudden change can be surprising and can disturb a system, particularly human economic and social systems. If it is only temporary change and the conditions return to the perceived "normal," the surprise may only have been troubling, not debilitating and destructive. Permanent change to a new condition, as some conceive of climate change, requires adaptation and the fomentation of resilience. Purveyors of this expectation think that they will only need to spend more money to cool their homes or that they will have to pay more for water because of the warming and drying of their weather patterns. But if the change to a new condition is only temporary and that change continues to move away from normal, perhaps even in nonlinear ways, then adaptation will need to be constant and resilience may be impossible. Conditions in the future may become unbearable and places may become uninhabitable because not only is climate changing, but so too are all of the other systems that are connected to and influenced by climates, like the changes that we continuously make to catch up to the new circumstances. What if the directions of the shifting conditions are not predictable? What if the changes seem to be chaotic, with rates of change constantly changing, and perhaps with retrograde shifts in some of the systems? The possibilities may become vast and unfathomable. Which adaptations will enable safety and survival? The information may be overwhelming, but systems-thinking can provide methods to understand how decisions about response to change can be made [16].

3.3 Ways not to respond to change

As soils are products of and are imprinted by the climate conditions of the geological past – it may take hundreds, if not thousands, of years to form a few centimeters of topsoil. A sudden shift in the climate at a location can create mismatches between weather regimes and soils. To survive, a farmer could shift to a different crop that fits the new meteorological reality, but that might have devastating economic financial impacts and there may be practical challenges [17]. Relocating the production of a specific agricultural crop to chase the "moving" climate may be a possibility if the land is accessible and free for the taking; this is how wildlife species occasionally respond to such changes [18]. But it is likely more difficult than one might imagine, particularly if the new location, with the right temperature and moisture regime, has inadequate soils (produced under dramatically different past climates) or soils that necessitate significant quantities of artificial inputs to grow the crop. Following the climate may be very difficult and unrealistic. Land ownership is an impediment to relocation of farming practices. Legal boundaries, political boundaries, social and cultural boundaries, and the logistical, infrastructural, and economic consequences and challenges in starting over, may prove to be insurmountable barriers. The choice may be to simply remain in place and change crops (i.e., adjust) or relocate and find ways to carry on (i.e., adapt). It is important to stress that even wild plants and animals cannot simply "move" and adapt with the changes [19]. There is often too much in their way, too.

Clearly, climate change would cause major interruptions in food production, globally. Completely giving in to the change and finding new ways to survive (because one rationally disposes of the luxuries and desires one had and accepts that it is most important to meet basic needs to survive) may be the most realistic use of human ingenuity to cope with new and evolving conditions if we choose not to mitigate the causes. This is an example of developing resilience. It used to be only the impoverished, landless, powerless, and desperate people who have lived this way for decades and centuries, but it may become reality for even those who were not impoverished, owned land, had power, and were not living in desperation [20]. Resilience at its most profound is the condition of learning to accept change by becoming immune to the threats, perhaps by simplifying life—coming to understand the difference between need and want, to have the fewest vulnerabilities possible, to survive by bending with the wind and flowing with the current, whichever directions they take you, to keep your head above the water until you find solid ground again.

Change may affect everybody and may impact every aspect of every life. The change to which one may be responding (like global warming-induced conditions, for instance) will also be impacting communities, societies, governments, other states of the world. There may be diminishing availability of assistance from friends, countrymen, strangers. The governments upon whom many depend for help may lack the resources to provide it. The structures of societies may begin to crumble. It may be every family, man, woman, and child for themselves as starvation, famine, poverty, dislocation, insecurity, and devolution ensue. An extreme future is no longer just science fiction. We are catching a glimpse of such a future in 2020 as the novel coronavirus "COVID-19" spreads around the world, disrupting normality. Perhaps it is a symptom of the changes that we have wrought.

An alternative may be to prevent what can be prevented before the changes become inevitable. We may preempt them so that they might not occur or will be muted. Instead of waiting for the forewarned changes, we could strive to understand their causes, and if possible, attenuate them. The prospects of forced adjustment, adaptation, and resilience may be so bleak and worrisome that the alternative, changing behaviors now to promote less future change, may be the preferred discomfort.

4. Climate change

So, global warming is happening and one of its consequences is that climates are changing. A warming Earth atmosphere-hydrosphere-biosphere system is not necessarily warming all climates. All of Earth's climates are not changing in the exactly the same ways (some may be warming, some drying, others becoming more hospitable to plant growth with longer growing seasons) or in the same directions (some may also be cooling, getting wetter, and seeing shorter growing seasons). And the pace of change is not consistent across space or time. Some are warming and some (particularly polar and high-latitude climates) are warming faster than others. Warming in Earth's polar regions is melting tundra permafrost [21], Arctic Ocean ice (and the Greenland ice sheet), and Antarctica's ice sheet. Antarctica's ice cap is shrinking in extent, it is thinning, and even hollowing out. The meltwaters from both poles are contributing to changes in ocean circulations, cooling of northward flowing warm currents like the Gulf Stream, furthering consequential shifts of atmospheric circulations from latitudinal to longitudinal flows in the northern hemisphere in the north Atlantic and in the Pacific [22]. The prospects of a colder Europe, caused by the diminishing flow of heat from the tropics, is troubling, given their regional agricultural production and dependence on regional production in many parts of the continent.

The belt of tropical climate (the intertropical convergence zone) is projected to expand due to increasing temperatures along the equator. This may intensify rainfall in the tropical rainforests and with added heat may widen the belt of rainforests and rainfall northward, southward, and upslope to higher elevations in mountainous areas, enabling spread of tropical weather conditions (hot and humid) that will last longer throughout the year [23]. The belt of low pressure will still migrate northward and southward with the progression of the astronomical seasons.

The expanded tropics may yield more and stronger tropical and subtropical storms (hurricanes, cyclones, typhoons), although the linkage of the changing "habits" of weather events to global warming is very tenuous [24, 25], primarily because weather is not climate. But this was one of the expected outcomes of global warming and changing climates expressed very early on in the discussion of the consequences of global warming as our experiences with subtropical storms is empirical (not theoretical), after all. We have not seen a rise in the number of storms annually, but the annual proportion of storms that are stronger does seem to be increasing. And another bit of a surprise to scientists, however, was a change in the storms themselves: they seem to be moving across Earth's surface (i.e., forward motion) more slowly, leading to longer and more devastating localized lashing by hurricane-force winds and extraordinarily heavy amounts of rainfall due to its increasing duration [26]. Both flooding and wind destruction have been increasing in the regions directly impacted by the storms. Warmer oceans strengthen and feed energy to extratropical storms, but they also swell (as water increases volume when it gets hotter and when it freezes) and inundate coastlines, adding to storm-surge problems on top of the flooding rains and high winds. We are finding that our present preparedness and planning for extratropical storms is being exceeded by the evolving nature of the storms. Future losses will be greater (and will spread farther into previously "safe" areas) and future costs for preparation, mitigation, and prevention of disasters will skyrocket alongside.

Sea-level rise is another problem on top of the complication of extratropical systems. Greater inundation extending further inland also "poisons" soils and water supplies with salty ocean water, causing additional problems, particularly for salt-intolerant vegetation (and people). Ocean water is also acidifying due to absorption of carbon dioxide (one of the three or four most important greenhouse gases) to create carbolic acid. Acidification is making ocean habitats less hospitable for ocean life like coral (another tangential consequence of global warming by greenhouse-gas loading of the atmosphere) [27]. Fish are one of the fundamental dietary proteins of people around the world. But fish stocks are declining, biodiversity is declining, and ocean-oriented coastal economies are declining globally. Some aquatic and terrestrial species will find ways to survive through biological adaptation, others may not [28]. But these consequences are technically not caused by changing climates, *per se*. They are warming-induced and are usually beyond the discussions had about "climate change." But as one might imagine, these are but a few of the changes produced by the problems derived from global warming.

5. Conclusion

Public discussions about contemporary human-induced global warming and climate change began to emerge in the early 1980s. Now, 40 years later, the discourse is mired in a polarized "debate," where those who desire action to counteract global warming and to prepare for changing climates are countered by a powerful minority of "deniers" and "skeptics" who refuse to even discuss the matter, because, they believe that the "problem" is not real (And even if it is, it is insignificant, because it is natural!) [29]. It is vital that we converse about the matter. Doing so is very fruitful [30]. It is crucial that the public (and many of the people who inform and "educate" the public) be inculcated with a deeper and clearer understanding of the concepts "climate," "change," and "climate change," by carefully, consistently, and meaningfully establishing standard definitions of these ideas and using them correctly all of the time. The implications of not understanding and not coming to terms with the threats we face are profound. The beliefs that we can empirically know "climate," that it refers to the entirety of "Earth's weather," that the "change" that might happen is either temporary or simply a one-and-done shift to a new steady-state condition, and that "climate change" means that it will be warmer everywhere on Earth in the coming decades, predetermine nonbelievers' responses, if they bother responding to global warming at all. At best, they expect that we may eventually need more air conditioning or assume we can just "move things" to more suitable places. Misunderstanding the basic terms of the problem begets a misunderstanding and a misinterpretation of the science behind the problem. Dismissing experts as elitist, calling global warming and climate change a hoax perpetrated by "the left" to destroy the economy, private wealth, freedom, and the world order, conservatives and libertarians push for collective social myopia and business as usual. Ironically, such an approach to the problems of global warming and changing climates will produce those very results.

Author details

John P. Tiefenbacher
Texas State University, San Marcos, Texas, United States of America

*Address all correspondence to: tief@txstate.edu

References

[1] Neukom R, Steiger N, Gómez-Navarro JJ, Wang J, Werner JP. No evidence for globally coherent warm and cold periods over the preindustrial common era. Nature. 2019;**571**:550-554. DOI: 10.1038/s41586-019-1401-2

[2] Crowley TJ. Causes of climate change over the past 1000 years. Science. 2000;**289**(5477):270-277. DOI: 10.1126/science.289.5477.270

[3] Vitousek PM. Beyond global warming—Ecology and global change. Ecology. 1994;**75**(7):1861-1876. DOI: 10.2307/1941591

[4] Dillon ME, Wang G, Huey RB. Global metabolic impacts of recent climate warming. Nature. 2010;**467**(7316):704-U88. DOI: 10.1038/nature09407

[5] Walther GR, Post E, Convey P, Menzel A, Parmesan C, Beebee TJC, et al. Ecological responses to recent climate change. Nature. 2002;**416**(6879):389-395. DOI: 10.1038/416389a

[6] Stamm KR, Clark F, Eblacas PR. Mass communication and public understanding of environmental problems: The case of global warming. Public Understanding of Science. 2000;**9**(3):219-237. DOI: 10.1088/0963-6625/9/3/302

[7] Whitmarsh L. What's in a name? Commonalities and differences in public understanding of 'climate change' and 'global warming'. Public Understanding of Science. 2009;**18**(4):401-420. DOI: 10.1177/0963662506073088

[8] Lehman B, Thompson J, Davis S, Carlson JM. Affective images of climate change. Frontiers in Psychology. 2019;**10**(960):1-10. DOI: 10.3389/fpsyg.2019.00960

[9] Sterman JD, Sweeney LB. Understanding public complacency about climate change: Adults' mental models of climate change violate conservation of matter. Climatic Change. 2007;**80**(3-4):213-238. DOI: 10.1007/s10584-006-9107-5

[10] Boykoff MT, Boykoff JM. Climate change and journalistic norms: A case-study of US mass-media coverage. Geoforum. 2007;**38**(6):1190-1204. DOI: 10.1016/j.geoforum.2007.01.008

[11] Carvalho A. Ideological cultures and media discourses on scientific knowledge: Re-reading news on climate change. Public Understanding of Science. 2007;**16**(2):223-243. DOI: 10.1177/0963662506066775

[12] Pfleiderer P, Schleussner CF, Kornhuber K, Coumou D. Summer weather becomes more persistent in a 2 degrees C world. Nature Climate Change. 2019;**9**(9):666-671. DOI: 10.1038/s41558-019-0555-0

[13] Sultan B, Defrance D, Iizmui T. Evidence of crop production losses in West Africa due to historical global warming in two crop models. Scientific Reports. 2019;**9**(12834):1-15. DOI: 10.1038/s41598-019-49167-0

[14] Lewontin RC. The meaning of stability. Brookhaven Symposium in Biology. 1969;**22**:13-23

[15] Wan XR, Jiang GS, Yan C, He FL, Wen RS, Gu JY, et al. Historical records reveal the distinctive associations of human disturbance and extreme climate change with local extinction of mammals. Proceedings of The National Academy of Sciences of the United States of America. 2019;**116**(38):19001-19008. DOI: 10.1073/pnas.1818019116

[16] Ballew MT, Goldberg MH, Rosenthal SA, Gustafson A, Leiserowitz A. Systems thinking as a pathway to global warming beliefs and attitudes through

an ecological worldview. Proceedings of the National Academy of Sciences of the United States of America. 2019;**116**(17):8214-8219. DOI: 10.1073/pnas.1819310116

[17] Khan Y, Bin Q, Hassan T. The impact of climate changes on agriculture export trade in Pakistan: Evidence from time-series analysis. Growth and Change. 2019;**50**(4):1568-1589. DOI: 10.1111/grow.12333

[18] Hofman-Kamińska E, Bocherens H, Drucker DG, Fyfe RM, Gumiński W, Makowiecki D, et al. Adapt or die—Response of large herbivores to environmental changes in Europe during the Holocene. Global Change Biology. 2019;**25**(9):2915-2930. DOI: 10.1111/gcb.14733

[19] Radchuk V, Reed T, Teplitsky C, van de Pol M, Charmantier A, Hassall C, et al. Adaptive responses of animals to climate change are Most likely insufficient. Nature Comunications. 2019;**10**(3109):1-14. DOI: 10.1038/s41467-019-10924-4

[20] Baarsch F, Granadillos JR, Hare W, Knaus M, Krapp M, Schaeffer M, et al. The impact of climate change on incomes and convergence in Africa. World Development. 2020;**126**(104699):1-13. DOI: 10.1016/j.worlddev.2019.104699

[21] Plaza C, Pegoraro E, Bracho R, Celis G, Crummer KG, Hutchings JA, et al. Direct observation of permafrost degradation and rapid soil carbon loss in tundra. Nature Geoscience. 2019;**12**(8):627-631. DOI: 10.1038/s41561-019-0387-6

[22] Zappa G. Regional climate impacts of future changes in the mid-latitude atmospheric circulation: A storyline view. Current Climate Change Reports. 2019;**5**(4):358-371. DOI: 10.1007/s40641-019-00146-7

[23] Colwell RK, Brehm G, Cardelus CL, Gilman AC, Longino JT. Global warming, elevational range shifts, and lowland biotic attrition in the wet tropics. Science. 2008;**322**(5899):258-261. DOI: 10.1126/science.1162547

[24] Coumou D, Rahmstorf S. A decade of weather extremes. Nature Climate Change. 2012;**2**(7):491-496. DOI: 10.1038/NCLIMATE1452

[25] Cutler MJ, Marlon J, Howe P, Leiserowitz A. Is global warming affecting the weather? Evidence for increased attribution beliefs among coastal versus inland US residents. Environmental Sociology. 2020;**6**(1):6-18. DOI: 10.1080/23251042.2019.1690725

[26] Chan KTF. Are global tropical cyclones moving slower in a warming climate? Environmental Research Letters. 2019;**14**(10):1040151-9. DOI: 10.1088/1748-9326/ab4031

[27] Pandolfi JM, Connolly SR, Marshall DJ, Cohen AL. Projecting coral reef futures under global warming and ocean acidification. Science. 2011;**333**(6041):418-422. DOI: 10.1126/science.1204794

[28] Diniz-Filho JAF, Bini LM. Will life find a way out? Evolutionary rescue and Darwinian adaptation to climate change. Perspectives in Ecology and Conservation. 2019;**17**(3):117-121. DOI: 10.1016/j.pecon.2019.06.001

[29] Weber EU, Stern PC. Public understanding of climate change in the United States. American Psychologist. 2011;**66**(4):315-328. DOI: 10.1037/a0023253

[30] Goldberg MH, van der Linden S, Maibach E, Leiserowitz A. Discussing global warming leads to greater acceptance of climate science. Proceedings of the National Academy of Sciences of the United States of America. 2019;**116**(30):14804-14805. DOI: 10.1073/pnas.1906589116

Chapter 2

The Developing World's Contribution to Global Warming and the Resulting Consequences of Climate Change in These Regions: A Nigerian Case Study

Angela Oyilieze Akanwa and Ngozi Joe-Ikechebelu

Abstract

Hundreds of millions of urban dwellers in low- and middle-income nations are at risk as 4-5 of the global weather-driven disasters experienced are consequent of a changing climate. Studies have shown that residents in least developed countries have ten times more chances of being affected by these climate disasters than those in wealthy countries. Further, critical views have it, that it would take over 100years for lower income countries to attain the resiliency of developed countries. Unfortunately, global South is surrounded by a myriad of socio-economic and environmental factors limiting their fight against climate crisis. It is this germane reality that provoked the cause of this review. Hence, this paper reviewed the developing world's contribution to global warming and the resulting consequences of climate change with focus on Nigeria. This purposive approach adopted an analysis of secondary data related to climate information. The findings from the paper affirmed that impacts of climate change in developing countries include loss in agriculture/forestry resources, water shortage, food insecurity, biodiversity loss, health risks among others. Finally, it identified the major factors that exacerbate climate crisis, the human actions that trigger global warming and adaptive and mitigation approaches to minimize climate change related disasters.

Keywords: climate change, human actions, environment, acclimatization options, GHG emission and Nigeria

1. Introduction

Humans have emitted about 450 billion tonnes of carbon since the industrial revolution which has contributed to the world's present climate crisis [1]. Additionally, the dependence on agro-economy, use of fossil fuels and industrial activities by developing countries have made huge contributions to increased levels of greenhouse gases (GHG) that have escalated global warming and sponsored a changing climate [2–5]. The United Nations Framework Convention on Climate Change [6] defined climate change as a change that is distributed directly or indirectly to human activity, altering the composition of the global atmosphere.

The changes in climate characteristics that include temperature, humidity, rainfall and wind, among others, are influenced by natural and human processes over long periods of time [7–10].

The principal and most abundant GHGs in the atmosphere are water vapour, carbon dioxide (CO_2), methane (CH_4), nitrous oxide (N_2O) and fluorinated gases (HFCs, PFCs, SF_6) [11–13]. Carbon dioxide (CO_2) makes up 60% of the greenhouse gases which alter the carbon cycle balance [14, 15]. They are emitted into our fragile atmosphere through human activities such as industrialization, burning of fossils, gas flaring, urbanization and agriculture. They increase global temperatures, thereby disrupting the current socio-economic and environmental processes [16, 17].

Human activities can also reduce the amount of carbon absorbed from the atmosphere through deforestation, land-use change, water pollution and agricultural production [18, 19]. In addition, developing countries are also involved in massive deforestation due to resource exploitation, urban expansion [2, 20] and agriculture in particular, which can cause carbon to be released from the soil at a faster rate than it is replaced.

Garrett Hardin's [21] concept of the "tragedy of the commons" identified that human unguarded actions are responsible for the depletion of natural resources and huge environmental compromise over the years. Rural areas that represent the resource base often get worse off by resource depletion and trickle-down effect of the urban vices. The aggressive demand placed on the environment and the unprotected means of processing raw materials are detrimental to climate change and the environment. Factors like improper/uncoordinated, non-participatory planning, non-implementation and enforcement of policies, among others, have been the bane of developing countries like Nigeria [2].

However, with the recent trend in global warming and the extent of man's responsibility in contributing to climate change, there is environmental concern towards sustainability. The concept of sustainability has become a major development process that will facilitate resource management and the minimization of impacts from human activities. The World Commission on Environment and Development [22] defines sustainable development as "development that meets the needs of the present generation without compromising the ability of future generations to meet their own needs".

The core component of the definition is the conservation and development of resources. Hence, there must be responsible interaction with the environment in the process of development to avoid depletion of natural resources and also allow for long-term restoration of environmental quality [3]. Unarguably, balancing human means of resource exploitation/production and environmental sustainability is a necessity since their outcomes have negative impacts on the environment. Hence, the need to review the extent of the developing world's contribution to global warming and the resulting consequences of climate change in these regions with focus on Nigeria as a case study and also to suggest possible solutions is the motivation for this study.

2. Background

Nigeria is ranked seventh in the world and has an estimated population of over 200.96 million. It accounts for about 47% of West Africa population including the largest populations of youth in the world [23]. Indeed, Nigeria is the giant of Africa with a federation that consists of 36 autonomous states, occupying a geographical area of 923,768 sq. km, has abundance of resources, is the largest oil exporter and is the largest natural gas reserve in the continent and yet a multi-ethnic and culturally diverse society with over 500 different ethnic groups [24]. Nigeria is blessed with physical, natural and human resources; however, she is surrounded by numerous environmental problems that have been exacerbated by climate crisis affecting its socio-economic and political sectors.

Nigerian and most developing countries in spite of their huge deposits in natural resources are still heavily dependent on agricultural production for their livelihoods. It is estimated that a total of 133 billion tonnes of carbon has been lost since humans first settled into agricultural life around 12,000 years ago. Also, crop production and cattle grazing have contributed almost equally to global losses too [25]. Further, a study carried out by Brandt et al. [26] also confirmed the contribution of agriculture to atmospheric carbon absorption by revealing that sub-Saharan Africa's vegetation carbon stocks changed between 2010 and 2016 with an overall loss of 2.6 billion tonnes of CO_2 in the past 7 years and yearly losses averaging at 367 million tonnes of CO_2. Developing countries like Ghana, Ivory Coast, Nigeria, Uganda and Kenya are experiencing the largest drop in carbon stocks with implications that climate change could make extreme events more frequent [26].

Hence, developing countries are exposed to climate vulnerability; as a result livelihoods, urban areas, infrastructures, human lives, health, socio-economic life and the built environment are threatened, particularly in Africa where sustainable resilient practices are mediocre. With low institutional capacity, sophisticated technology, finance and adequate resources are limited in these poor and most vulnerable regions [27].

It is quite alarming that the rate at which temperature increases is accelerating global warming incidence, revealing the need to operate environmentally friendly options and adaptive measures for climate regulation in order to ameliorate drastic weather predictions [3, 4, 28]. It is the need to review the developing world's contribution to global warming and the consequences of climate change in these regions with focus on Nigeria as a case study that has prompted these pertinent and their possible solutions:

1. What is the present climate situation in Nigeria and other developing countries?

2. What are the major limiting factors that exacerbate climate risks in these regions?

3. What are the impacts of climate change in these regions?

4. What efforts are these regions making to salvage the climate situation?

5. What are the research gap(s) in climate change studies in Nigeria and other developing countries?

3. Method

The search of databases was carried out by the authors. The search of articles was limited to the English language only. The search list included studies, grey literature and policy papers with relevant documents. The information sources included articles from Springer, PubMed, ResearchGate, Google Scholar and Social Sciences Abstracts from January 1990 to June 2019. Referenced sections of the articles identified were used to find additional references not retrieved by the initial search engines. The following search terms were employed (text words and climate related subject headings): climate change or climate crisis and sustainable development, health impacts, tragedy of commons concept and community-based approaches to climate change in developing world (in low- and middle-income nations) and climate change in Nigeria. The search terms were reviewed and tested with an information specialist. Review articles were adopted using measures across climate change challenges, sustainable development concept and community-based approaches in developing and developed nations. All titles and abstracts were screened by AA, and the decision to include or exclude was recorded by both authors. Studies were managed by using a reference management software. The reviews particularly related to climate change in high-income countries were excluded. However, the focus of the study was on low- and middle-income nations in Asia and Africa with more emphasis on a highly populated West African sub-region (Nigeria). Data were extracted from papers on climate change and health impacts. Data was also extracted from related recent studies carried by the authors, and the study took a narrative synthesis of results in line with research objectives.

3.1 Global warming and climate situation in the developing world

Developed countries have been able to minimize the adverse effects of climate change due to some factors such as natural advantage, high adaptation techniques, high technology, mechanized agricultural system and wealth status. For example, countries such as Norway, New Zealand, Sweden, Finland and Denmark were found to be most prepared to adapt to climate change [29, 30].

However, a report by ND-GAIN [31] indicated that it will take more than 100 years for the world's poorest countries to reach the current adaptive capacity of higher-income OECD countries. Unfortunately, developing countries such as Nigeria have had major setbacks that have escalated climate change problems. Given the predictions made by climate change research organizations and scientific reports, there are increased frequency and intensity of climate change in developing countries, especially in parts of Asia and Africa. This vulnerability has led to several environmental impacts observed majorly as flooding, drought, food shortages, heat waves and health risks, among others [17]. **Table 1** summarizes these environmental impacts.

The future of developing countries in the face of a changing climate if appropriate actions are not quickly enforced will definitely be unbearable. Indeed, sub-Saharan Africa will be the most vulnerable region with poorly supplied infrastructure, the domestic per capita food production has declined by 10% in the last 20 years, and about 800 million people are poorly fed [33, 34]. Overtly, the prevailing crisis covering poor living and feeding conditions will be escalated further by the consequences of changing climate with prevalent issues of food insecurity and health risks.

Further, Podesta and Ogden [35] asserted that West Africa suffers the greatest losses due to climate change, with increases ranging between 36 and 44% of the losses for the entire continent and between 42 and 60% of agricultural regional GDP. IPCC [36] and Climate Change [37] predicted that glacier melt will increase flooding and rock avalanches and affect water resources in Tibet, India and Bangladesh. Bangladesh will be most affected because it both is low-lying and has about 13 million people and Dhaka's GDP per capita is the lowest of all the cities. This will affect its capacity to adapt to climate change [17].

UN-Habitat's [17] report further stated that more than a billion people could have water shortages by the 2050s. Southeast Asia, especially the heavily populated mega delta regions, will be at risk from flooding. Around 30% of Asia's coral reefs are likely to be lost in the next 30 years due to multiple stresses and climate change.

No	Environmental impacts	Socio-economic resources and sectors affected
1	Changes in rainfall patterns	Water resource settlements:
	Increased frequency and severity of: Floods Droughts Storms Heat waves Changes in growing seasons and regions	Agriculture and forestry Food security Human health Infrastructure (e.g. transport) Displacement of inhabitants and loss of livelihood
2	Changes in water quality and quantity	Coastal management
3	Sea level rise	Industry and energy
4	Glacial melt	Disaster response and recovery plans

Adapted from UK Parliament Publications [32].

Table 1.
Climate change impacts in the developing countries.

Damage to coastal cities will impact on tourism too. For instance, the coastal city of Mombasa, in Kenya, could lose 17% of its land, which will affect amenities and features that draw tourism. Changes in rainfall will increase diarrheal diseases mainly associated with floods and droughts and possibly increase malaria distribution [38].

By 2080, an increase of 5–8% of arid and semi-arid land is projected under a range of scenarios. Already by 2020, between 75 and 250 million people will be exposed to increased water stress due to climate change. Agricultural production, including access to food, is projected to be severely compromised. In some countries, yields from rain-fed agriculture could be reduced by up to 50%. Sea level rise will affect major cities in low-lying coastal areas, such as Alexandria, Cairo, Lomé, Cotonou, Lagos and Massawa [38]; Climate Change 2007 and [36]. Hence, developing countries are in dire need of sustainable measures for establishing a low-carbon region.

3.2 Factors responsible for climate change vulnerability in developing countries

Several studies and authors have provided a number of interrelated factors which are responsible for climate change vulnerability in developing countries. They are categorized as natural, human and inadequate infrastructures, poor urban planning, low level of adaptation capacity and inadequate preparedness [39–42]. UN-Habitat [17] reported that developing countries have lower adaptive capacity covering human, financial and other resources to adapt to the effects of climate change.

Natural factors have left most developing countries vulnerable such as the Democratic Republic of the Congo, Central African Republic, Eritrea, Burundi and Chad which are affected by climate change [31]. People living in vulnerable areas like deltas and semi-arid and coastal regions are equally vulnerable. More than 1 million people live in deltas, semi-arid lands glacier- and snowpack-dependent river basins in Africa and Asia, and the hot spot regions that are the most vulnerable to climate change [43]. **Table 2** showed some developing countries and their vulnerability and capability ranking in response to climate change. It showed that Uganda has the highest vulnerability index (160) and the lowest readiness capacity (36.9) to climate change.

Further, adequate housing and its associated facilities also contribute to the vulnerability of developing countries to climate change. According to UN-Habitat [17], around 30% of the urban population in developing regions was living in the slums in 2012, and this figure was over 60% in sub-Saharan Africa. Kampala in Uganda has been experiencing rapid urbanization and slum expansion as over 50%

Country	Vulnerability		Readiness		Overall	
	World rank	Score	World rank	Score	World rank	Score
Bangladesh	140	0.534	148	0.327	140	39.7
India	118	0.473	122	0.377	120	45.2
Nepal	128	0.495	115	0.393	122	44.9
Pakistan	115	0.469	142	0.341	126	43.6
Tajikistan	78	0.409	131	0.357	111	47.4
Burkina Faso	145	0.555	155	0.319	148	38.2
Ghana	124	0.484	102	0.442	108	47.9
Mali	164	0.604	138	0.348	156	37.2
Senegal	146	0.556	127	0.368	137	40.6
Ethiopia	144	0.553	146	0.330	145	38.9
Kenya	147	0.557	159	0.312	154	37.7
Tanzania	143	0.550	144	0.353	139	40.1
Uganda	156	0.573	159	0.312	160	36.9
Botswana	123	0.483	76	0.494	94	50.5
Namibia	141	0.547	99	0.445	122	44.9

Source: University of Notre Dame Global Adaptation Index [29].

Table 2.
Showing the measure of countries' vulnerability and their capability to respond to climate change.

of its urban population live in slums [44]. These types of settlements are often built in areas that are more vulnerable to the effects of climate change because they are less expensive and poorly built with little capacity to resist events such as flooding.

In addition, developing countries have experienced growth that has been largely unplanned, uncontrolled and lacking adequate infrastructure, as most of these cities emerged without master plans [45]. Indeed, human and urban planning factors such as growth in population, poor governance, decaying infrastructure and lack of proper environmental planning and management exacerbated extreme weather incidents in Nigerian largest cities [46].

Okonkwo and Akanwa [47] affirmed that the underlying major causes that exacerbated flooding in urbanized cities such as Lagos, Kano, Ibadan and Anambra is the lack of sustainable urban designs complicated by improper refuse disposal, erecting of structures on flood plains, improper development of urban planning and infrastructure and other indiscriminate actions that interfere directly or indirectly with the free flow of water.

Table 3 showed African countries and their level of urban access to these basic infrastructures: sanitation, safe water and health. **Table 3** indicated that Benin had to be the least urban access to basic infrastructures, while South Africa had 99% in 2000 and Namibia had 100% in 2000 for sanitation and safe water infrastructures, respectively.

3.3 Global warming and climate change: the Nigerian case

The world population has continued to gallop coupled with urbanization trends [50] and has exploded from an estimated maximum of 15 million people in prehistory to the 7 billion humans today. About 70% of the world's urban population lived

Country	Sanitation		Safe water		Health
	1990%	2000%	1900%	2000%	1990–2000%
Benin	46	46	NA	74	42
Burkina Faso	88	88	74	84	NA
Cameroun	99	99	76	82	NA
Comoros	98	98	97	98	NA
Cote d'Ivoire	78	NA	89	90	NA
Ghana	59	62	83	87	25
Guinea	94	94	72	72	25
Lesotho	NA	93	NA	98	NA
Madagascar	70	70	85	85	NA
Namibia	84	96	98	100	NA
Nigeria	77	85	78	81	67
Senegal	86	94	90	92	40
South Africa	NA	99	NA	92	NA
Togo	71	69	82	85	NA
Tanzania	97	98	80	80	93

Source: World Bank [48] African Development Indicators. NA = not available (adapted: [49]).

Table 3.
Percentage of urban population with access to some basic infrastructure in selected countries in sub-Saharan Africa.

in developing countries in 2010 [51]. Increasing and large population numbers with land densities make people aggressively dependent on available natural resources and the environment for survival, especially in developing countries like [52].

In 2017, Nigeria had a population of 190 million with 2.6% annual growth and GDP capita of 1.969 USD and ranked 139th globally [53]. Nigeria is projected to contribute 10% of the 2.2 billion increases in global population expected by 2050 [53]. The economy of Nigeria is mainly dependent on crude oil resources and rain-fed agriculture can have a huge influence on ecosystems and temperature trends and climate change impacts.

There has been provoking evidence of climate change in Nigeria with increase in the temperature trend since 1901. A study carried out by Odjugo [19] investigated the mean air temperature in Nigeria for 105 years covering between 1901 and 2005. He reported that the temperature increase for 105 years was 1.1°C and the mean air temperature in Nigeria was 26.6°C. This showed higher than the global mean temperature increase of 0.74°C recorded since 1860 which is the original date of inception for scientific temperature measurement [36, 54]. Also, NIMET [55] confirmed that there have been changes in Nigeria's climate as proven observations from 1941 to 2000 showed evidence of long-term temperature increase in most of the country. The most significant increases were recorded in the extreme northeast, extreme northwest and extreme southwest, where average temperatures rose by 1.4–1.9°C. The rising temperature trend showed an increase in temperature when averaged over the country from about 26.2°C in 1951 to about 27.6°C over the years.

Obviously, increasing temperature levels have been ignited by human activities globally and in Nigeria as well. There are several human actions that contribute to increased temperatures in Nigeria. With huge deposits of crude oil in the Niger Delta region, Nigeria accounts for roughly one sixth of worldwide gas-flaring nations and flares about 75% of her gas [56]. Other livelihood patterns such as the clearing of forestland for firewood and wood charcoal, agriculture and commercial logging that have largely contributed to land-use emissions which account for 52% of our GHG production, with recent estimates showing that Nigeria is responsible for 490 million tonnes of GHG emissions (CO_2 equivalent) annually, just over 1% of global production. About 39% of this arises from land-use change and forestry, 33% from energy production (oil and gas extraction and the power sector), 14% from waste (incineration of municipal waste), 13% from agriculture and 2% from industry [5].

Further, apart from activities that emit GHG in Nigeria, there are also human actions that reduce carbon sinks; such as deforestation and agricultural production. They are means by which huge amounts of vegetation cover are lost. Satellite data has revealed that between 1987 and 2011, lowland, mangrove and freshwater forest areas in the Niger Delta have decreased by 15–40% [57]. Another satellite data and GIS study carried out by Akanwa et al. [2] affirmed that 402.855 ha of green cover have been lost in Ebonyi State, Nigeria, as a result of quarrying activities. The resultant loss of carbon sinks has been accompanied by significant loss in biodiversity and the ecosystem services they offer, both of which exacerbate global warming. Notably, if the present situation is unrestrained in Nigeria, there may be a crisis of temperature increase between the middle (2.5°C) and high (4.5°C) in the year 2100 [19].

The undeniable temperature increase in Nigeria has incited extreme weather consequences and impacts like increase in rainfall intensity, flooding, displacement of people, destruction of buildings, infrastructures, loss of lives, biodiversity loss, health risks, overflooding of farmlands and food shortages. In fact, there has been abnormal rainfall pattern and corresponding extreme events such as floods and drought since 1951–2015 [55]. Coastal cities in Nigeria like Port Harcourt, Yenagoa, Warri, Anambra and Calabar have experienced an increase in rainfall pattern observed by storm surges in recent times [33, 58, 59]. It was estimated that a metre rise in sea level will displace about 14 million people from the coastal areas of Nigeria [60].

Also, over 70% of Nigerian communities are agrarian, and their activities are rain controlled. Climate change which affects rainfall patterns and threatens livelihood with extreme weather events such as thunderstorms, heavy winds and floods could devastate farm lands, leading to crop failure [33]. Moreover, climate variations accentuate pest and crop disease spread which will potentially pose a threat to livestock and food security [61]. Notably, there have been some observable changes in agricultural production especially in the southern parts of Nigeria where farming activities normally start from March when the early wet season starts and farming normally kicks off. Recently, there are observed deviations in the commencement of agricultural activity, for instance, due to changes in the arrival of the wet season [55, 62, 63]. It is projected that Nigeria would experience yield decreases in cash crops of 5–40% by 2050 if no technological innovation in farming methods is implemented [33, 64, 65].

Nigeria has also experienced drought in the northern region whereby nomadic herders are moving southwards into the fertile Middle Belt to find suitable pastures for their livestock [26]. This has brought about ethnic clashes between the rural farmers and herdsmen. The shrinking of Lake Chad basin is indicative of climate change's footprint in the Sahel region. Odjugo and Ikhuoria [66] also observed that Nigeria North of 120 N is under severe threat of desert encroachment and sand dunes are now common features of desertification in states such as Yobe, Borno, Sokoto, Jigawa and Katsina. This has prompted massive emigration and resettlement of people to areas less threatened by desertification, thereby creating crisis elsewhere.

Notably, there are resultant health risks influenced by high temperatures and stagnant water from flooding. This leads to high incidence of the spread of malaria, water-borne diseases and the transmission of contagious diseases such as cholera and influenza. This has resulted in deaths, sickness and injuries due to increased exposure to heat waves [7, 67, 68].

The above stated impacts cuts across agriculture/forestry, water shortage, food security, biodiversity, infrastructure, human and animal life, drought, human health and livelihoods. Indeed, Indeed, climate change affects a vast, if not the entire sphere of our environment and human existence. Further, seven countries are predicted to suffer the largest average losses in the agricultural sector with Nigeria suffering the highest in the group [35].

Developing countries like Nigeria depend solely on natural resource harvesting and agriculture and these activities are easily affected by climate change. Considering that the presently, Nigeria's economy is affected due to these aforementioned impacts and future predictions continue to reveal a projected severe economic effect of climate change in Nigeria with a rise in sea level from 1990 level to 0.3 m by 2020 and 1 m by 2050 and rise in temperature of up to 3.2°C by 2050 following a drastic climate change [69]. Ultimately, climate change impacts could result in a loss in GDP of between 6 and 30% by 2050, worth an estimated US$100–460 billion dollars. By 2020, if no adaptation is implemented, between 2 and 11% of Nigeria's GDP could potentially be lost [5, 70, 71].

3.4 The Nigerian response to climate change

In addition, Nigeria adopted its Climate Change Policy Response and Strategy (CCPRS) in 2012 to ensure an effective national response to the multi-faceted impacts of climate change. The main goals of the CCPRS include implementation of mitigation measures that will promote low-carbon as well as sustainable and high economic growth; enhancement of national capacity to adapt to climate change; raising climate change-related science, technology and research; and many more. The National Adaptation Strategy and Plan of Action for Climate Change in Nigeria (NASPA-CCN) have been developed and describes the adaptation priorities. According to the NASPA [72] report, the programme will enable federal, state and local governments, civil society, the private sector and various agencies and institutions to effectively integrate climate change adaptation concern into their development policies and programmes such as water and other natural resource, agriculture, health and infrastructure. NASPA will give priority to community-level input as an important source of information, recognizing that grassroots communities are key stakeholders and providing a voice to the most vulnerable (including women and youth), ensuring that everyone is represented in the plan [46, 72]. The actualization of this strategy will depend heavily on our ability to respond to the prevailing climate change effects as a nation and our survival instinct to protect our future.

In conclusion, these strategies should be simple, strategic, concise and applicable so that these developments can be the foundation of technological and policy innovation that can minimize GHG and achieve a low-carbon growth in developing countries. Also, it is necessary to create maximum awareness of the total environment, share its concern and have a collective goal towards providing sustainable solutions and equitable developments. Nevertheless, the Global South should focus on the reality of adapting to climate change by finding ways to live with overflowing sea levels, scarce drinking water, higher peak temperatures, depleting species and agriculture altering weather patterns, health risks and poorer food production [73].

3.5 Mitigation and adaptation strategies for minimizing climate change in developing regions

UNFCCC have already identified two ways to address climate change: first through mitigation of climate change by reducing greenhouse gas emissions and enhancing sinks and secondly through adaptation to the impacts of climate change. Mitigation comprises all human activities aimed at reducing the emissions or enhancing sinks of greenhouse gases such as carbon dioxide, methane and nitrous oxide [74, 75]. **Table 4** summarizes areas that require mitigation and the corresponding activity-solutions to enhance the carbon sinks.

Adaptation in the context of climate change refers to any adjustments that take place in natural or human systems in response to actual or expected climatic stimuli or their effects or impacts, aimed at moderating harm or exploiting beneficial opportunities [7, 74].

Areas	Activities
Demand-side, brownfield energy efficiency	Commercial and residential sectors (buildings) Public services Agriculture Industry
Demand-side, greenfield energy efficiency	Construction of new buildings
Supply-side, brownfield energy efficiency	Transmission and distribution systems Power plants
Renewable energy	Electricity generation, greenfield projects Transmission systems, greenfield Heat production, greenfield or brownfield projects
Transport	Vehicle energy efficiency fleet retrofit Urban transport modal change Urban development Interurban transport and freight transport
Agriculture, forestry and land use	Afforestation and reforestation Reducing emissions from the deforestation or degradation of ecosystems Sustainable forest management Agriculture Livestock Biofuels
Waste and wastewater	Solid waste management Wastewater treatment Waste recycling
Non-energy GHG reductions	Industrial processes Air conditioning and cooling Fugitive emissions and carbon capture
Cross-sector activities	Policy and regulation Energy audits Supply chain Financing instruments Low-carbon technologies Activities with greenhouse gas accounting

Source: Ref. [76].

Table 4.
Typology of mitigation activities.

Vulnerability focus ← → Impacts focus

Theme 1	Theme 2	Theme 3	Theme 4
Addressing drivers of vulnerability	**Building response capacity**	**Managing climate risks Activities**	**Confronting climate change Activities**
Activities seek to reduce poverty and other non-climatic stressors that make people vulnerable: •Livelihood diversification •Literacy and education Women's rights • Community health • Food Security • Water supply ,• sanitation	Activities seek to build robust systems for problem solving: •Natural resources management • Weather data collection, forecasting •Disaster early warning systems •Communication systems	seek to incorporate climate information into decision-making: •Climate proofing projects •Disaster response planning •Drought-resistant crops; Cropping-systems •Robust, adaptive technologies	seek to address impacts associated exclusively with climate change: •Relocation due to sea level rise (SLR) •Coastal defenses from SLR Managing glacial-lake outburst floods (GLOF) • Extra storage to capture glacial melt

Traditional development funding → ← New and additional funding

Figure 1.
The continuum of adaptation activities from development to climate change vulnerability focus. Source: Adapted from Ref. [77].

McGray et al. [77] showed in their diagram the adaptation and development as an option for dealing with limitations faced by developing countries whereby their development needs or vulnerabilities can be dealt with in order, not to sabotage the adaptation techniques applied. **Figure 1** showed themes 1 and 2 and also showed that unmet development needs (as manifest in poverty or weak institutional capacities) drive vulnerability, including vulnerability to climate change. Such development needs need to be addressed to ascertain that adaptation activities (as in themes 3 and 4) can be effective and sustained. From this perspective, a separation of development from adaptation may in fact be counterproductive to the ultimate goal of achieving climate resilience.

Generally, developing countries and their cities have lower emissions, but there is limited data, and GHG emission levels vary from place to place. There are three sectors that give the highest GHG emissions; they are residential, commercial and transportation. Hence, developing countries can have good planning practices and climate-smart practices around these sectors [17]. They can implement adaptive strategies such as urban greening and farming in order to contribute to carbon sequestration [3, 4]. Urban vegetation, such as trees, can store carbon, provide shade for buildings and reduce air and noise pollution in cities. Urban farming also provides green spaces and carbon sequestration benefits by reducing the urban heat island effect. It can have a positive impact on food security and reduce urban poverty [3]. Green spaces such as roofs and parks bring ecological and social benefits to urban areas. Green roofs can improve drainage of rainwater and thus reduce rainwater run-off in case of extreme weather events [78–81].

Green spaces and public green areas, such as parks, improve the quality of the social life urban dwellers by creating integration and accessibility [82–88].

Also, developing countries can incorporate sustainable urban designs and construction by making use of robust materials for buildings or locating buildings at high or more stable grounds. These approaches could be ad hoc which involves taking opportunity of a crisis situation. For instance, in Mozambique in 2010, its city Maputo was severely flooded, and the "Living with Flood" initiative took up the project of building schools and community halls that would serve as shelter during future floods [17]. In Nigeria, for example, an energy efficient transport system can be introduced to minimize the high levels of carbon monoxide emitted into the atmosphere.

An ad hoc approach could require partnering with environment- or health-based NGOs to solve a problem. Nigeria has received support from the WHO on water, erosion and drainage projects, among others. Another example is the City of São Paulo; it entered into agreements with the Brazilian company Biogás. Biogás constructed facilities at two landfill sites for a total investment of US$ 90 million. At the Bandeirantes site, a system captures the methane gas and channels it into a combined heat and power plant. The two landfills together now generate 10% of the city's electricity requirements. To date, the credits generated by reduced emissions have yielded some 48 million Euros, which the city splits 50/50 with Biogás. The City Council of São Paulo has used its share of the revenues to develop parks and squares in the poor neighbourhoods surrounding these landfills [17]. The challenge was that the 10 million inhabitants generated 15,000 tonnes of garbage daily, and harnessing methane gas was an asset, while reducing GHG emission and improving livelihood were expedient.

Also, developing cities can have specific plans with an implementation strategy. For example, an action plan has been adapted by Cape Town in South Africa for energy and climate change, which has (11) objectives and targets. One of the targets is 10% renewable and cleaner energy supply in 2020. Another plan is to build more compact and resilient-efficient city. The implementation of the plan involved more than 115 projects. Mitigation approach could also be achieved through incorporating climate change into existing plans, policies and programmes, for example, transport, public health, energy management and disaster risk reduction plan. For example, Cambodia's coastal city of Preah Sihanouk is planning for climate change by mainstreaming into existing planning process [17].

Generally, most developing countries have begun to develop alternative policy frameworks, for example, through national adaptation programmes. These have focused on climate-proofing infrastructure projects, such as transport and irrigation systems, improved disaster monitoring and management and better land-use planning [89]. For instance, the Bangladesh Climate Change Strategy and Action, adopted by the government of Bangladesh in 2009, seeks to guide activities and programmes related to climate change in Bangladesh. The strategy contains 44 programmes formulated around six themes which include food security/social protection/health, comprehensive disaster management, mitigation/low-carbon development and capacity/institutional strengthening. Thirty-four programmes listed under five themes are wholly or partially focused on adaptation [48]. Adaptation measures are also incorporated into disaster preparedness in Bangladesh. Furthermore, Orindi and Murray [90] acknowledged the progress being made in East Africa on integrating adaptation into the most vulnerable sectors. This is also similar in other African countries such as Tanzania, Uganda and Sudan in their national communication to the UNFCCC.

The mitigation and adaptation strategies enumerated above surmises the UN climate speech of 2017, by the UN Climate Change Executive Secretary, Patricia Espinosa, She called for a more holistic view of health throughout the world, with an inclusion of the concept of planetary health, a re-establishment that there is an intricate connection of the health of humankind to the right condition of the overall environment and other living beings. The Rockefeller-Lancet commission defined planetary health as the health of human civilization and the state of the natural systems on which it depends [91]. Planetary health is a new and emerging field across disciplines intersecting physical/natural sciences and the health of human. In addition, planetary health is acceptable by international organization, the global north universities, countries and their NGO actors, due to the input of sustainable environmental interventions. However, there is need for the inclusion of developing nations in this new emerging concept. The impacts of climate change in developing nations reveals an ominous threat to the health of the citizenry and the health of the natural systems which they depend.

We have used and abused our natural resources with negative impacts to the populace. We see diseases, food insecurity, poverty, hunger and malnutrition with high risks of vitamin deficiencies. Hence a basic and introductory thrust on which planetary health lies is the innovative and transformative actions for integrative approach to further develop an evidence base to inform solutions that simultaneously address human health, environmental sustainability and economic development [92], in essence, one that directly connects human and animal health with the health of the planet [93] so as to improve the lives of individuals, families and communities in Nigeria and beyond

5. Conclusion

Generally, this study assessed the local and regional contributions of Nigeria and other developing countries to global warming and the resulting consequences of climate change. Findings from the study showed that the impacts of climate change in developing countries include loss in agriculture/forestry resources, water shortage, food insecurity, biodiversity loss, infrastructure damage, loss of human and animal life, drought and health and livelihoods risks. Also, certain factors that were highlighted such as rapid urbanization, poor urban design, inadequate infrastructure, inadequate meteorological information, poor awareness and low levels of literacy combined with human actions such as industrialization, gas flaring, burning of fossil fuels and agricultural practices have immensely escalate climate change-related disasters. These have affected the socio-economic sectors of Nigeria and other developing countries made towards adaptation, adjustment and resilience in Africa.

These issues of global warming are real, and the myriad of limitations faced by developing countries show huge potentials that may totally collapse the economic, social and environmental processes of developing countries particularly in the area of agriculture, tourism and natural resource if sustainable actions are not quickly implemented. Africa battles with data documentation and monitoring systems and the known levels of GHG emitted, moreover, the human health implications cannot be properly accounted for. Hence, Nigeria and other developing countries can become part of planetary health approaches and alliances for sustainable development and safe public health presently adopted in institutions of global North. As Planetary health looks at the health of human civilization and the state of the natural systems on which they depend (Seltenrich, [91]). Though an emerging, multidisciplinary, cross-sectoral concept within borders, planetary health reverts attention to the anthropogenic degradation of our planet with potential to sustainable environmental solutions for the climatic plague of humanity.

Finally, this study advocates an applicable mitigation and adaptation option that is tailored to each developing country's peculiar socio-economic and environmental challenges. The need for community based participatory approaches with population groups that are more disproportionately burdened from the impact of climate change is critical in Nigeria as well as developing countries. It is obvious that community academic partnerships that are wholly driven based on community needs are necessary at this time of climate emergencies in Nigeria. Though community driven partnerships are of high financial inputs, however, the impacts may come with health and social generated equities. We do know that climate change is an example of a complex problem with multi sectoral and interdisciplinary perspectives. Community oriented interventions have reached a critical level in developing countries especially Nigeria. Though one may argue about the level of built capacity of Nigeria academic in community-based research in climate change, the need for international organizations and universities to ally with local communities and academic can bring on positive outcomes for Nigeria. This can be achieved through ad hoc, strategic research and mainstreaming actions and into plans, projects and programmes on different sectors such as proffering of landscape planning, greening, urban farming, environmental with health interventions and awareness and implementation of holistic protective laws and policies as effective approaches. It also suggests that developing nations can take advantage of their situation and stand together by forming networks to act on climate change. They can make emissions reduction commitments, evaluate adaptation strategies and advocate for national and international financial support from research evidence. This way developing nations can adapt to climate change comfortably and create a resilient environment and climate-proof socio-economic systems.

Author details

Angela Oyilieze Akanwa[1*] and Ngozi Joe-Ikechebelu[2]

1 Department of Environmental Management, Chukwuemeka Odumegwu Ojukwu University, (COOU), Uli Campus, Anambra State, Nigeria

2 Department of Community Medicine/Primary Health Care, Chukwuemeka Odumegwu University, Amaku-Awka Campus, Anambra State, Nigeria

*Address all correspondence to: angela.akanwa1@gmail.com

References

[1] Sanderman J, Hengl T, Gregory J, Fiske G. Soil carbon debt of 12,000 years of human land use. Proceedings of the National Academy of Sciences of the United States of America. 2017;**114**:9575-9580

[2] Akanwa AO, Okeke FI, Nnodu VC, Iortyom ET. Quarrying and its effect on vegetation cover for a sustainable development using high resolution satellite image and GIS. Journal of Environment and Earth Science. 2017;**76**(4):1-12

[3] Akanwa AO, Mba HC, Ogbuene EB, Nwachukwu MU, Anukwonke CC. Potential of agroforestry and environmental greening for climate change minimization. In: Abhishek R et al., editors. Climate Change Impact and Agroforestry System. UK: International Standard, CRC-Apple Academic Press and Taylor and Francis; 2019. p. 389. ISBN: 9781771888226

[4] Akanwa AO, Mba HC, Jiburum U, Ogboi KC. Strategies for combating climate change. In: Jhariya MK et al., editors. Sustainable Agriculture Forest and Environmental Management. International Standard Book. Springer: Nature Singapore Pte Ltd; 2019. pp. 2-47. ISBN: 978-981-13-6829-5

[5] Ogbonna CU. Adaptation on climate change in developing countries. In: A Need in the Delta Niger Region of Nigeria; 2014. Available from: http//:www.researchgate.net/publisher300275441

[6] UNFCCC. United Nations Climate Change Conference. COP 21 or CMP 11 was held in Paris, France, from 30th November to 12th December; 2015

[7] IPCC. Impact, adaptation and vulnerability. In: Contribution of Working Group II of the Intergovernmental Panel on Climate Change to the Third Assessment Report of IPCC; London: Cambridge University Press; 2001

[8] Masika R, editor. Gender, Development and Climate Change. Oxford: Oxfam; 2002

[9] McMichael AJ, Lindgren E. Climate change: Present and future risk to health, and necessary response. Internal Medicine. 2011;**270**(5):401-413

[10] Nnaji CE. Effects of climate change on household water sources in Benue State, Nigeria. In: An MSc. Project Thesis Submitted in the Department of Agricultural Extension; University of Nigeria: Nsukka; 2012

[11] National Climatic Data Center. National oceanic atmosphere administration. In: Land and Ocean Temperature Departure; 2015

[12] National Research Council. Advancing the science of climate change. In: U.S. Environmental Protection Agency (EPA); 2011

[13] World Meteorological Organisation. Greenhouse gas concentrations in atmosphere reach new record. 2013

[14] Food and Agriculture Organization (FAO). Agriculture, forestry and other land use emissions by sources and removals by sinks 1990-2011 analysis. Climate energy and tenure division, FAO Statistics Division; Working Paper ESS/14-02; 2006

[15] Greenhouse Protocol for Cities. Global Protocol for Community-Scale Greenhouse Gas Emission Inventories. IISD Report Services. Washington, DC, USA: 2013. Available from: https://ghgprotocol.org/greenhouse-gas-protocol-accounting-reporting-standard-cities

[16] Onyekuru NA, Marchant RA. Nigeria's response to the impacts of climate change. Developing resilient and ethical adaptations. Journal of Agricultural and Enviornmental Ethics. 2011;**25**(4)

[17] UN-Habitat. Cities and climate change: Global report on human settlements 2011. In: Greenhouse Gas Bulletin; 2011. p. 3

[18] Akanwa AO, Onwuemesi FE, Chukwurah GO, Officha MC. Effect of open cast quarrying technique on vegetation cover and the environment in south-eastern, Nigeria. American Scientific Research Journal for Engineering, Technology and Sciences (ASRJETS). 2016;**21**(1):227-240

[19] Odjugo AO. General overview of climate change impacts in Nigeria. Journal of Human Ecology. 2010;**29**(1):47-55

[20] Akanwa AO, Ikegbunam FI. Adverse effects of unregulated aggregate Exploitationin south-eastern Nigeria. EPRA International Journal of Research and Development (IJRD). 2017;**2**(3):167-177

[21] Hardin G. The tragedy of the commons. Journal of Science. 1968;**162**:1243-1248

[22] World Commission on Environment and Development (WCED). Our Common Future. Oxford: Oxford University Press; 1987

[23] Nigerian Population Review. World population prospects 2019. 2019. Available from: http:// worldpopulationreview.com

[24] Hughes JB, Daily GC, Ehrlich PR. Population diversity: Its extent and extinction. Science. 1997;**278**:689-692

[25] Brandt M. Human population growth offsets climate-driven increase in woody vegetation in sub-Saharan Africa. Nature Ecology and Evolution. 2017;**1**:0081

[26] Brandt M, Jean-Pierre W, Chave J, Tagesson T, Penuelas J, Ciais P, et al. Satellite passive microwaves reveal recent climate-induced carbon losses in African dry lands. Nature Ecology and Evolution. 2018;**2**:827-835

[27] UNFCCC. Report of the Conference of Parties on the Thirteenth Session. Bali, Indonesia: United Nations Framework convention on Climate Change, Geneva; 2007

[28] Nishioka S, Harasawa H, editors. Global Warming: The Potential Impact on Japan. Tokyo: Springer; 1998. pp. 3-8

[29] ND-GAIN. Global Adaptation Index of Developed Countries and Adaptation Capacity. University of Notre Dame; 2014

[30] Ogbuabor JE, Egwuchukwu EI. The impact of climate change on the Nigerian economy. International Journal of Energy Economics and Policy. 2017;**2**:217-223

[31] ND-GAIN. Global Adaptation Index: Measure of Country Vulnerability and Capability to Respond to Climate Change. University of Notre Dame; 2013

[32] UK Parliament Publication. Climate Change: International Framework for Action. Printed in the UK for the Stationery Office Limited On behalf of the controller of Her Majesty's Stationery Office. 185031 03/06 19798. Published by TSO (The Stationery Office). 2006. Available from: www.tsoshop.co.uk

[33] Akanwa A, Ezeomedo IC. Changing climate and the effect of gully erosion on Akpo community farmers in Anambra state. Journal of Ecology and Natural Resources. 2018;**2**(6):1-12

[34] Slater R, Peskett L, Ludi E, Brown D. Climate change, agricultural policy and poverty reduction—How much do we know? Natural Resource Perspectives. London: Overseas Development Institute (ODI). 2007;**109**:2-6

[35] Podesta J, Ogden P. The security implications of climate change. The Washington Quarterly. 2008;**31**(1):115-138

[36] Intergovernmental Panel on Climate (IPCC) 2007. Climate change 2007. In: The Fourth Assessment Report (AR4). Available from: http://www.ipcc.ch/pdf/assessment-report/ar4/syr/ar4_syr_spm.pdf [Accessed: 10 August 2009]

[37] Climate Change. Impacts, Adaptation and Vulnerability. Contribution of Working Group II to the IPCC'S Fourth Assessment Report (AR4) (9780521 88010) Hardback. 2007

[38] UNEP. The environmental food crisis: The environment's role in averting future food crises. In: A UNEP Rapid Response Assessment; United Nations Publications; 2009

[39] Agboola T. The prospect for private sector involvement in Urban Management functions in Nigeria. Review of Urban and Regional Development Studies. 1995;**6**:135-149

[40] Fadamiro JA. Outdoor spaces and their landscape qualities: A comparative analysis of three neighbour hoods in Lagos, Nigeria. Journal of Urban and Environmental Research. 2000;**2**:55-67

[41] George CK. Attaining sustainable development in lagos metropolis: A challenge for town planners. In: Paper Presented At 27th Annual Conference of Nigeria; Institute of Town Planners: Benin-City, Edo State; 1999

[42] George CK. The Challenges of Urbanization in Nigerian Urban Centres: The Lagos Mega City Situation: A Town Planner's Perspective. Lagos: Libro-Gem Books Ltd; 2001

[43] Collaborative Adaptation Research Initiative in Africa and Asia-CARIAA. Targeting Climate change Hot Spots in 15 Countries. In: Providing a Baseline picture of national adaptation policy and practice in developing countries; 2018. Available from: http://www.cariaa.net

[44] Mabasi T. Assessing the vulnerability, mitigation and adaptation to climate change in Kampala City. In: Paper Presented at the Fifth Urban Research Symposium, Cities and Climate Change: Responding to an Urgent Agenda, 28-30 June; Marseille: France; 2009

[45] Wahab B. Ibadan: A rapidly growing city in need of a master plan. In: Paper Presented at the Architects' Congress/Annual General Meeting Organized by the Nigerian Institute of Architects; Oyo State Chapter 7 July 2011; Ibadan; 2011

[46] Building Nigeria Response to Climate Change-BNRCC. Local action on climate change: BNRCC's Quarterly News Letter October 2010. BNRCC: Ibadan, Nigeria; 2010. pp. 1-12

[47] Okonkwo AU, Akanwa AO. Sustainable Urban design: A tool for reducing flood vulnerability and building resilient cities in Nigeria. On transforming Nigeria through the sustainable development goals: Options for environmental management. In: A Paper Presented at the Annual Conference for Environmental Management Association of Nigeria (EMAN); Nigeria; 2017

[48] World Bank. Economics of adaptation to climate change: Bangladesh. 2010. pp. 1-86. Available from: http://climatechange.worldbank.org/sites/default/files/documents/EACC_Bangladesh.pdf [Accessed: 09 November 2011]

[49] Ibem EO. Community-led infrastructure Provision in Low-Income Urban Communities in Developing Countries: A Study on Ohafia, Nigeria. Cities. 2009;**26**(3):125-132

[50] Cohen JE. World Population in 2050: Assessing the Projections. USA: Rockefeller University and Columbia University; 2002

[51] UN. United Nations Population Division. 1999

[52] Nigeria Environmental Study Action Team (NEST). Local action on climate change: BNRCC's Quarterly Newsletter. 2010. pp. 1-15

[53] UNDP. Human development data (1990-2017). In: United Nation Development Programme Report; 2017

[54] Spore. Climate change. Special issue. 2008. p. 23

[55] Nigerian Meteorological Agency (NIMET). 2018 Seasonal Rainfall Predictions. In: A Report by NIMET; 2018. p. 47

[56] World Bank. A Report on Development and Climate Change: A Strategic Framework for the World Bank Group-Technical Report. 2008. Available from: http://popenknowledge.worldbank.org/handle/10986/28201

[57] Ayanlade A, Drake N. Forest loss in different ecological zones of the Niger Delta, Nigeria: Evidence from remote sensing. Geo Journal. 2016

[58] Odjugo PAO. An analysis of rainfall pattern in Nigeria. Global Journal of Environmental Sciences. 2005;**4**(2):139-145

[59] Odjugo PAO. The impact of climate change on water resources: Global and regional analysis. The Indonesian Journal of Geography. 2007;**39**:23-41

[60] Abu B. Sea level rise and the Niger Delta of Nigeria. Journal of Wetland. 2007;**3**(1):44-52

[61] Onyenechere E, Igbozurike U. Women crop farmers' adaptation to rainfall variability and climate change in Amaraku community, Imo state, Nigeria. In: Nwajiuba C, editor. Climate Change and Adaptation in Nigeria. Weikersheim: Margraf; 2008. pp. 65-72

[62] Onyenechere E. Climate change and spatial planning concerns in Nigeria: Remedial measure for more effective response. Journal of Human Ecology. 2010;**32**(3):137-148

[63] Munonye J, Okoli V, Nwajiuba CU. Rainfall variability and arable crop production in the humid south of Nigeria. In: Nwajiuba C, editor. Climate Change and Adaptation in Nigeria. Weikersheim: Margraf; 2008. pp. 45-52

[64] Tack J, Lingenfelser J, Jagadish SV. Disaggregating sorghum yield reductions under warming scenarios exposes narrow genetic diversity in US breeding programs. Proceedings of the National Academy of Sciences of the United States of America. 2017;**114**(35):9296-9930

[65] Zougmore R, Partey S, Ouédraogo M, Omitoyin B, Thomas T, Ayantunde A, et al. Toward climate-smart agriculture in West Africa: A review of climate change impacts, adaptation strategies and policy developments for the livestock, fishery and crop production sectors. Agriculture and Food Security. 2016;**5**(26)

[66] Odjugo PAO, Ikhuoria AI. The impact of climate change and anthropogenic factors on desertification in the semi-arid region of Nigeria. Global Journal of Environmental Sciences. 2003;**2**(2):118-126

[67] Duru P, Ume N. Climate change and the incidence of malaria in AhiazuMbaise, Imo state.

In: Nwajiuba C, editor. Climate Change and Adaptation in Nigeria. Weikersheim: Margraf; 2008. pp. 95-100

[68] World Health Organisation. Global climate change: Implications for international public health policy. Bulletin of the World Health Organization. 2007;**85**(3):161-244

[69] IPCC. Climate change 2014: Mitigation of climate change. In: Contribution of Working Group III to the Fifth Assessment Report of the Intergovernmental Panel on Climate Change. Cambridge, United Kingdom, New York, USA: Cambridge University Press; 2014

[70] DFID. Impact of climate change on nigerian's economy. In: Final Report, British Department for International Development (DFID); UK; 2009

[71] FMEAN. National environmental, economic and development study (NEEDs) for climate change in Nigeria. In: Federal Ministry of Environment Abuja, Nigeria (Special Climate Change Unit); 2010. pp. 1-20

[72] NASPA. Developing Nigeria's climate change adaptation response. In: Nigeria's 2010 National Adaptation Strategy and Plan of Action (NASPA); 2010. pp. 1-9. Available from: http://www.naspanigeria.org/docs/NASPA_BkgrndDoc.pdf [Accessed: 11 November 2011]

[73] Stanley I. Nigeria and climate change adaptation. In: International Society of Sustainability Professionals Insight Report May 2012; 2012

[74] Klein JTR, Schipper ELF, Suraje D. Mitigation and adaptation into climate and development policy: Three research questions. Environmental Science & Policy. 2005;**8**:579-588

[75] UNFCCC. Fact sheet: Need for mitigation. 2009. pp 1-7. Available from: http://unfccc.int/files/press/backgrounders/application/pdf/press_factsh_mitigation.pdf [Accessed: 07 November 2011]

[76] Joint MDB Report on Mitigation Finance. A Report by a group of Multilateral development Banks. 2011. Available from: http://www.eib.org/attachments/documents/joint-mdb-report-on-mitigation-finance-2011pdf

[77] McGray H et al. Weathering the Storm: Options for Framing Adaptation and Development. Washington DC: World Resources Institute; 2007

[78] Jay W, Scott B. Essential Environmental. Pearson: Glenview; 2011

[79] De bon H, Parrot L, Moustier P. Sustainable urban agriculture in developing countries. A review. Agronomy for Sustainable Development. 2010;**30**(1):21-32

[80] Mkwambisi DD, Fraser EDG, Dougill AJ. Urban agriculture and poverty reduction: Evaluating how food production in cities contributes to food security, employment and income in Malawi. Journal of International Development. 2011;**23**(2):181-203

[81] Zezza A, Tasciotti L. Urban agriculture, poverty, and food security: Empirical evidence from a samples of developing countries. Journal of Food Policy. 2010;**35**(4):265-273

[82] David B. What is landscape design. 2017. Available from: www.thespruce.com [Accessed: April 23]

[83] Gontier M, Mortberg U, Balfors B. Comparing GIS-based habitat models for application in EIA and SEA. Environmental Impact Assessment Review. 2010;**30**:8-18

[84] Ibimilua AF. Key issues on landscape planning in the context of

environmental sustainability. European Scientific Journal. 2014;**10**:143-156

[85] Lundy L, Wade R. Integrating sciences to sustain urban ecosystem services. Physical Geography. 2011;**35**:653-669

[86] Owonaiye O. Landscape Planning as a Means of Controlling Erosion: A case study of Oke-Ogba layout in Akure, Ondo State. 2012. Available from: www.academia.edu [Accessed: 23 April 2018]

[87] Santra SC. Environmental Science. London: New Central Book Agency Ltd; 2005

[88] Zheng B, Zhang Y, Chen J. Preference to home landscape: Wildness or neatness? Landscape and Urban Planning. 2011;**99**:1-8

[89] UN-DESA. World Economic and Social Survey 2009: Promoting Development, Saving the Planet. United Nations, New York; 2009. pp. 1-56

[90] Orindi V, Murray L. Adapting to Climate Change in East Africa: A Strategic Approach, Gatekeeper Series 117. London: IIED; 2005. pp. 1-16

[91] Seltenrich N. Down to Earth: The energy field of planetary health. Environmental Health Perspectives. 2018;**126**(10):109001

[92] Osofsky SA, Ponsgiri MJ. Operationalizing planetary health as a game-changing paradigm. Health impact assessment are key. The Lancet Planet Health. 2018;**2**(2):e54-e55

[93] Espinosa V. Human resources for health in Ecuador's new model of care. Pan American Journal of Public Publication. 2017;**41**:e96

Chapter 3

Deciphering the Climate Change Conundrum in Zimbabwe: An Exposition

Nelson Chanza and Veronica Gundu-Jakarasi

Abstract

The notion that climate change has created development opportunities largely remains poorly understood despite phenomenal evidence that points toward positive gains across the broad socio-economic spectrum. Current understanding has largely concentrated on the negative effects of climate change, with limited exposition on the benefits associated with climatic responses. This article collates and reviews evidence that interventions to curtail climate change impacts have unlocked several development opportunities and potentially contribute in improving the living standards of many communities in Zimbabwe. It argues that although climate change effects permeate all the socio-economic development sectors of the country, the collective interventions by government, development partners and individuals on mitigation and adaptation actions could lead to a development trajectory that is evident in a number of indicators toward poverty alleviation, particularly through improved food, energy, water, and health access. The article, however, questions the sustainability of these unfolding benefits and advises on the need to enhance mechanisms for climatic programming in the country's development plans, policies and strategies.

Keywords: adaptation, climate change, clean energy, food security, health, mitigation, poverty, water access

1. Introduction

The phenomenon of climate change has been nothing short of spectacular. Recent scholarship confirms earlier evidence that change and variability in the climate system, primarily triggered by anthropogenic greenhouse gas (GHG) emissions, will have far reaching global consequences [1–5]. The events associated with climatic phenomena, largely noticeably as extreme temperatures, storms, droughts and floods, are said to be more frequent and severe in developing countries [1, 3]. The reasons for such a regional risk divide in exposure to climate change are beyond climatic. The Intergovernmental Panel on Climate Change (IPCC) captures them as non-climatic drivers of vulnerability, which summarily include poor governance, conflicts and instabilities, inequalities, hunger, poverty and disease [2, 6, 7]. Zimbabwe is not an exception to the climatic disturbances. The major climatic issues are evidenced by declining water resources, fall in agricultural productivity, biodiversity decline, geographical spread of vector-borne diseases and pestiferous nature of problem pests, and volatile weather and climatic disasters [8–10].

In its Fifth Assessment Report (AR5), the IPCC identifies challenges and opportunities in both mitigation and adaptation responses [6]. Generically understood as a proactive measure to prevent or minimise harm, mitigation in climate science and practice carries a dual meaning. From the distinction given by the IPCC as mitigation of disaster risk and disaster and mitigation of climate change through reducing GHG emissions and enhancing carbon sinks, both definitional strands are beneficial to societies practising climatic interventions, albeit with some challenges. Drawing from the conceptual scheme adopted by Wilbanks and Sathaye [11], which classifies mitigation as structural (technological) and unstructural (economic structure, societal organisation and individual behaviour), a number of opportunities could be unlocked, particularly in addressing the negative effects of climate change. On the other hand, the definitional scope of adaptation extends from the mitigation of disaster risk and disaster strand, primarily focusing on actions taken to respond to climatic events. As such, it is critical to evaluate what climatic responses can do to communities intended to benefit from those policy or strategy systems. The purpose of climate mitigation, therefore, is to stabilise the climatic system and lessen pressure on adaptation. Several scholars (for example, see [11, 12]) have pointed out the complementary roles of mitigation and adaptation, arguing that adaptation is difficult or even a futile if mitigation fails to minimise the magnitude of the costs to be handled. This argument tends to shape current climate policy regimes, the recent one being the Paris Agreement, which aims to "Strengthen the global response to the threat of climate change, in the context of sustainable development and efforts to eradicate poverty..." [13].

The dividends of such a global policy framework, where they exist, remain largely obscured by the attention to the magnitude of observed and anticipated climatic threats and, therefore, become poorly understood. Africa is one of the regions seriously affected by climate change. The reasons for this are reported by Boko et al. [14] and later reinforced by Niang et al. [15] as relating to other factors as unequal access to resources, enhanced food insecurity and poor health management systems, which exacerbate the vulnerabilities of many communities in the region. Despite low levels of adaptive capacity reported by Klein et al. [16], adaptation success stories associated with higher adaptive capacities have been noticed in some countries mainly in North Africa [17]. Overall, within the continent, individual, household or micro level adaptive capacities are shaped by functional institutions, access to assets and collective action [15, 18]. These opportunities enhance the ability of people to make informed decisions in exploiting the beneficial aspects of responding to climate change.

Realising the effects that climate change poses to its broad socio-economic development sectors, Zimbabwe has not been complacent in responding to climate change, albeit experiencing challenges. Current evidence of climate change in Zimbabwe portrays a predominantly challenging situation. Existing knowledge on the sectoral impacts of climate change in the agriculture, water, energy, industry and health sector point to a negative state (for example, see [8, 9, 19]). This article challenges this confinement by broadening the focus to examine opportunities associated with the phenomena of climate change. In this article, the effects of climate change are deciphered. To do this, the article adopts a sector-based analysis to show how the various socio-economic development sectors are experiencing climate change. The policy and institutional field is also evaluated to understand the supporting system for climate change responses. This background forms the basis for understanding the climatic interventions that have been made or planned for each sector identified. Within this exposition, it is shown that both challenges and opportunities exist. Given the limited scholarship treatment on the latter, the article draws from empirical evidence which shows that several opportunities have been unlocked and/or remain open for beneficial exploitation by individuals, communities and institutions as they take advantage of the climate change response agenda. The article underscores the need to take continuous stock of achievements made in the country's development sectors as mitigation and adaptation interventions gather momentum. The sustainability question of these observed and anticipated benefits is given considerable examination throughout the article.

2. Research methodology

The article largely adopts an empirical methodological approach, which is inarguably more appropriate to exposit the climatic experiences in Zimbabwe. This has been complemented by case study reviews to deepen the empirical analysis on local evidence of climatic interventions and to draw from experiences of similar climate change-induced responses from other countries mainly drawn from Africa. Within this approach, the study utilised a combination of sector-based and policy and institutional analytic frameworks to evaluate existing climatic practices that the country has embraced since 1992, when the country started to participate in global climate change regimes. Essentially, climate change responses in Zimbabwe can be traced since 1992, the period that recorded a major milestone in embracing climatic responses following the signing and ratification of the United Nations Framework Convention on Climate Change (UNFCCC). Given the evident influence of non-climatic factors of vulnerability that are reported by the IPCC [1], the analysis is broadened to incorporate the development policies that govern the broad socio-economic sectors in Zimbabwe. Thus, the empirical evidence presented in this discussion is largely drawn from existing official government reports, including other documents such as national budget statements. The utilisation of official documents might have marginalised some climatic activities and statistics outside the mainstream government records, particularly those not captured by the Climate Change Management Department (CCMD). However, given the coordinating role of climate change responses by the CCMD, the study was able to capture the official facts and statistics about climatic phenomena and response interventions in Zimbabwe. Where some figures required to be updated, particularly on funds that were received to implement climate change projects, officials from the CCMD were engaged to verify the recorded statistics.

The reports and documents reviewed were also mainly sector-specific, although at the level of analysis, some overlaps exist among the agriculture, water, energy and health sectors that are discussed. Sectoral analysis was used both to deepen the analysis of climatic impacts in each sector and to examine the adequacy of the current climatic practices. The policy and institutional review also informed the state of the support systems in place to tackle the climate change challenge. It was necessary to assess if the available institutions are adequately capacitated to drive the climate change response agenda, particularly in a context where the sustainability of climatic interventions is increasingly getting some attention in global policy regimes of climate change and disaster risk management [13, 20].

3. Sectoral impacts of climate change in Zimbabwe

The evidence of climate change in Zimbabwe can best be presented by adopting a sector-based analysis, as noted earlier. This section presents observed and anticipated impacts of climate change in the agriculture, water, energy and health sectors, which are the most representative and highly vulnerable to climate change.

3.1 The agriculture sector

With reference to climate change impacts, one of the sectors that has drawn research, policy and practical attention is the agricultural sector. Climatic events such as extreme temperatures, increase in frequency of extreme weather events, and rainfall variability are projected to affect agriculture in many ways. Noticeable impacts are already being felt in increased crop failures, pests, crop disease, and the degradation of land and water resources [8, 9]. The role of agriculture as an economic enabler deserves emphasis. Agriculture promotes value chain systems and contributes about 60% to manufacturing, while consuming almost 40% of the industrial output. The sector also has a share of around 30% of export earnings, constitutes 60–70% of employment, and about 19% of GDP [21]. In this way, the sector provides a major source of livelihood for over 70% of the country's population [21]. Owing to its deep intermesh with the rest of the economy, disruption in agriculture from climatic shocks could lead to overall economic decline. Clearly, this is a de-coupling challenge that needs appropriate interventions by taking advantage of the climate change situation.

More than 70% of crop farming practice is rain fed [22]. This suggests that agriculture, food security, and nutrition are all highly sensitive to changes in rainfall associated with climate change. Specifically, climate change has been observed to trigger shifts in agricultural farming regions, with consequential loss in productivity [23, 24]. Given the regional differentiation of the climate system, where productivity follows the agro-ecological zones, climate change is believed to cause shrinkage in the highly productive regions. Agricultural performance productivity generally shows an east-west productivity gradient mainly influenced by the rainfall and temperature. This scheme, however, has been dismissed as obsolete and largely misleading in representing the current farming and ecological regions [23]. The main documented reasons for threats in farming production are high temperatures and precipitation irregularities reported by Mutasa [25] and Unganai [26]. The situation is blamed for causing arid environments that make it difficult for most food and cash crops to grow. The crops that are highly sensitive to heat include maize (a staple crop), tobacco (the major cash crop), wheat, soya beans, among others. Studies have shown a suitability gradient of different crops under different climatic scenarios. The areas suitable for maize production are projected to decrease by 2080, while spatial suitability of crops such as cotton and wheat is expected to increase by the same year [9]. However, it is believed that the north central and eastern areas of the country will likely to be less vulnerable to support production of common crops such as maize, sorghum and cotton [9, 27].

In Zimbabwe, climate change also impacts heavily on livestock. Generally, evidence of climate adaptation in the agriculture sector is moving towards livestock production as a drought tolerant practice [10]. However, as shall be discussed in the next section, there are indications of limits to using livestock as a strategy to adapt to climate change that are pointed out by Tubiello et al. [28] and Chanza [10]. This is because the decline in plant productivity associated with arid environments will likely affect rangelands and feed. The direct impacts of changes in temperature and water scarcity on animals are expected to constrain adaptation efforts. Though not well documented and understood, the indirect effects are likely to be through increased pests and diseases of livestock and decline in pasture yield. The cattle population is estimated to be about 5.5 million. Instead of increasing by over 2% per annum, the national cattle herd has been facing climatic threats. For example, the drought experienced in 2014/2015 and 2015/2016 seasons is believed to have aggravated the foot and mouth disease. The disease rapidly spread as cattle moved wider in search of water and forage and was reported in six of the country's ten provinces [21]. This affected commercial activities involving cattle and other livestock products.

It is also important to point out that the climatic impacts explained here are not uniformly experienced across the country's tenure systems. Communal and small-scale farmers are more likely to be negatively affected by the warming temperature and variability in rainfall [8, 27]. As detailed in the next subsection, the situation also impacts heavily on food security particularly to small-scale subsistence farmers whose operations are not covered by irrigation schemes.

3.2 The water sector

The total amount of water available for the country is estimated to be about 20 million megaliters of freshwater [9]. It is critical to point out that the availability of this water is largely climatic [29]. Replenishment of the water is through rainfall leading to runoff into streams, rivers, dams and lakes. Some of it collects into vleis and surface depressions or ends up as ground water stores in the form of aquifers. The country has an estimated dam population of over 8000 [8]. Zimbabwe also has seven river catchment areas, namely Mazowe, Manyame, Save, Runde, Sanyati, Gwayi and Umzingwane. The sensitivity of these catchments to climate change varies with their location and with the type of land use practices in the catchments. The 2080 model predictions generally show a significant reduction in surface water resources. The areas to the north eastern and the eastern of Zimbabwe are projected to have a surplus in surface water. However, the western and southern parts of Zimbabwe, where Umzingwane, Runde, Gwayi and Save are located, are projected to experience significant decrease in runoff and desiccation of the catchments [9, 30].

Increased water scarcity associated with climate change can also be seen in depreciation in ground water levels. The common understanding is that water tables are becoming deeper. Where communities used to easily access water through shallow wells, they now need to dig deeper to tap up the water [10]. This is clear evidence that the groundwater is getting depleted owing to a drier climate. A report by the IPCC [31] confirms that rural communities relying on low-cost dug wells and boreholes are now exposed to serious water stress owing to interruptions in recharges resulting from drought.

3.3 The energy sector

Climatic concerns in the energy sector are twofold. The sector is not only a driver of climate change due to GHG emissions, but is also affected by its impacts [32]. Given that the sector drives other socio-economic factors, such impacts need to be carefully examined. Currently, the country is not producing enough energy to meet demand and it covers the deficit through electricity imports. In rural areas, there are immense challenges facing attempts to extend the national grid. Energy deficits are high in the rural areas with an estimated 19% of the rural people only having access to reliable electricity. Without electricity, farmers cannot process their crops, add value or diversify their livelihoods thereby affecting agricultural productivity. In schools and homes, children struggle to study without light and are cut off from modern technology thus affecting education performance. Health institutions are also not spared from intermittent power cuts and this affect the national health delivery system [33].

Hydro-power contributes a significant proportion to the country's electricity generation. Recurrent drought in the past few years coupled with changing rainfall patterns within the southern African region have led to the decrease in water levels of major reservoirs [34]. A conspicuous impact of climate change affecting the energy sector has been isolated in the 2015/2016 season. The water levels in Zimbabwe's main lake, Lake Kariba, dropped to below 30%. This situation seriously affected power generation in the country. Similarly, in Kenya, droughts that occurred between 1999 and 2002 drastically affected hydro-power generation, falling by 25% in 2000. The resultant cumulative loss in generation was variously estimated at between 1.0 and 1.5% of total GDP. These negative climate impacts have affected other sectors of freshwater distribution and food production [35].

Zimbabwe uses a mix of energy sources. These include fossil fuels (coal, coal bed methane and imported petroleum) and clean energy sources (hydropower, biofuel and solar). The sector faces challenges from rising population and economic demands. Climate change is also expected to exacerbate the energy supply situation. The energy sector constitutes about 49% share of total GHG emissions in CO_2 equivalent [36]. However, as shall be discussed later, there are also opportunities created by climate change in the sector.

3.4 The health sector

Evidence suggests that climate change will affect human health in various ways. Africa is already experiencing high burdens of health outcomes whose frequency, magnitude and spatial range is anticipated to grow [15]. These challenges, largely triggered by temperature and precipitation extremes, manifest in malnutrition, diarrheal diseases, and malaria and other vector-borne diseases. Climate change is also expected to exacerbate the human exposure to heat waves and direct exposure to ozone owing to elevation of ozone in the troposphere [37, 38]. There is a gender dimension to these problems, with evident disproportionate impacts on women, children and people living with disabilities [39, 40].

In Zimbabwe, observed health burdens of climate disturbances largely emanate from high frequencies and severity of floods, storms and droughts, including geographic spread of infectious disease vectors. The geographical range of malaria and other mosquito-borne diseases, such as dengue; increases in the problem of diarrheal diseases, and of water-borne pathogens such as cholera and typhoid, are worrisome [22, 41]. Hartmann et al. [41], using sixteen climate change scenarios, reveal that the geographical distribution of malaria could change, with previously unsuitable areas becoming suitable for transmission as the ecology of vectors and pathogens is altered. Matawa and Murwira [42] also projected expansion in habitats of certain disease vectors owing to changes in temperature and rainfall in some parts of the country. There are also fears that disease epidemics in addition to other stressors such as food insecurity, chronic malnutrition, and HIV and AIDS are eroding the resilience of households, rendering them less resilient and more vulnerable to health problems. Although mainly attributed to water contamination, the recent outbreaks in cholera and typhoid could also be partly blamed on climate change. A case in point is the repeated outbreaks of cholera that recorded over 98,000 cases and more than 4000 deaths between August 2008 and June 2009 [43] and over 6500 cases and 31 deaths reported by 20 September 2018 [44]. The Cyclone Idai, which was downgraded to a tropical depression on the 16th of March 2019 caused high winds and heavy precipitation in Manicaland Province, riverine and flash flooding and subsequent deaths, destruction of livelihoods and properties, with Chimanimani and Chipinge districts being the most affected. The Ministry of Health and Child Care (MoHCC), with support from development partners, had to urgently move in to lead the health response, including preventing outbreaks of epidemic diseases such as cholera [45].

4. Policy and institutional responses to climate change

In line with the climate mitigation and adaptation agenda articulated in multilateral environment agreements (MEAs), Zimbabwe's policy space has largely been characterised by active participation in international environmental laws and subsequent ratification of the MEAs. Reviewing this policy space is necessary to show the main activities in the national policy regime and the reapable benefits thereof. Thus, the country's response to the three main climate based MEAs, namely the UNFCCC, the Kyoto Protocol and the Paris Agreement is assessed in this section. These instruments unlock opportunities in the form of knowledge and skills acquisition through training, technical assistance, technology transfer, funds received, materials or equipment accessible to the country.

Zimbabwe signed and ratified the UNFCCC in 1992. The coordination for the implementation of this Convention is done by the Climate Change Management Department (CCMD) in the Ministry of Lands, Agriculture Water, Climate and Rural Resettlement. The UNFCCC is supported by other important instruments namely, the Kyoto Protocol and the Paris Agreement. The purpose of the UNFCCC is to prevent dangerous human interference with the climate system. It covers climate change assessments, mitigation and adaptation. Focus of the Convention is on stabilising GHGs at a level to be achieved "... within a time-frame sufficient to allow ecosystems to adapt naturally to climate change, to ensure that food production is not threatened, and to enable economic development to proceed in a sustainable manner." The Kyoto Protocol is an international agreement linked to the UNFCCC, which commits its Parties by setting internationally binding emission reduction targets. Although drawing much attention to the developed countries as principally responsible for the current high levels of GHG emissions in the atmosphere as a result of industrial activity, Zimbabwe also ratified the Protocol in 2009 [9].

The Paris Agreement builds upon the UNFCCC and has managed, for the first time, to bring all nations into a common cause to undertake ambitious efforts to combat climate change and adapt to its effects, with enhanced support to assist developing countries to do so. Zimbabwe signed the Agreement on 22 April 2016, ratified it on 7 August 2017 and was entered into force on 6 September 2017. The Agreement's central aim is to strengthen the global response to the threat of climate change by keeping global temperature rise below 2°C above pre-industrial levels and to pursue efforts to limit the temperature increase even further to 1.5°C. Additionally, the agreement aims to strengthen the ability of countries to deal with the impacts of climate change. To reach these ambitious goals, appropriate financial flows, a proposed technology framework and an enhanced capacity building framework will support action by developing countries and the most vulnerable countries, in line with their own national objectives. The Agreement also provides for enhanced transparency of action and support through a more robust transparency framework. It requires all Parties to put forward their best efforts through nationally determined contributions (NDCs) and to strengthen these efforts in the years ahead. This includes requirements that all Parties report regularly on their emissions and on their implementation efforts [13, 46].

Under these instruments, Zimbabwe has developed the National Climate Change Response Strategy (NCCRS) in 2014; Intended Nationally Determined Contribution (INDC) in 2015; National Climate Policy (NCP) of 2017; the First, Second and Third National Communication to the UNFCCC. The country also conducted United Nations Programme on Reducing Emissions from Deforestation and Forest Degradation (UN REDD+) Capacity Needs Assessment. **Table 1** summarises the interventions made so far and the benefits that accrue to the country.

Through the Adaptation Fund that is established to finance concrete adaptation projects and programmes in developing countries that are vulnerable to the adverse effects of climate change, Zimbabwe is likely to benefit from this funding window. There is an on-going process for the Environmental Management Agency (EMA) to be accredited as a National Implementing Entity (NIE), to access the funds. The country also got support from the Common Market for East and Southern Africa (COMESA), UNDP, Global Water Partnership Southern Africa, UNECA, Climate

MEA	Focal point	National action plans/strategies	Source of funds	Funds received (US$)
UNFCCC	CCMD	NCCRS, 2014	COMESA, UNDP, Global Water Partnership Southern Africa, UNECA, Climate Technology Centre and Network (CTCN) & Environment Africa	100,000
UNFCCC	CCMD	Energy and Water Efficiency Audit for 10 selected pilot companies	Climate Technology Centre and Network (CTCN)	250,000
UNFCCC	CCMD	Climate Smart Agriculture Manual development	Climate Technology Centre and Network (CTCN)	100,000
UNFCCC	CCMD/EMA	Coping with Drought Project	Special Climate Change Fund (SCCF)	1,000,000
		Scaling Up Adaptation	Special Climate Change Fund (SCCF)	3,980,000
UNFCCC	CCMD	National Climate Policy (NCP)	Government of Zimbabwe, UNDP, Global Water Partnership Southern Africa, UNECA, Climate Technology Centre and Network (CTCN) & Environment Africa	130,000
UNFCCC	CCMD	Third National Communication to UNFCCC	UNEP	400,000
UNFCCC/Kyoto Protocol	CCMD/Forestry Commission	UN REDD+ Capacity Needs Assessment	UN-REDD	50,000
Paris Agreement	CCMD	Intended Nationally Determined Contribution (INDC), 2015	UNEP from Zimbabwe's GEF STAR Allocation and French Embassy in Zimbabwe	200,000
Paris Agreement	CCMD	Climate Change Technical Assistance-NDC MRV Framework development	World Bank	1,500,000
UNFCCC	CCMD/EMA	NIE Accreditation	Adaptation Fund/ South-South Cooperation	50,000
UNFCCC	CCMD	GCF Readiness Programme	Green Climate Fund	300,000

MEA	Focal point	National action plans/strategies	Source of funds	Funds received (US$)
UNFCCC	CCMD/EMA	National Adaptation Plan	Green Climate Fund	3,000,000

Table 1.
Interventions for climate related MEAs implementation (source: [47]).

Technology Centre and Network (CTCN) and Environment Africa. These provided support towards the development of the NCCRS, NCP, UNFCCC COP participation and other capacity building on climate change issues negotiations, including development of a Climate-Smart Agriculture (CSA) Manual, Technical Assistance on Climate Change readiness, and NDC MRV Framework development. So far, Zimbabwe has one Clean Development Mechanism (CDM) registered project known as the Sable Chemicals Tertiary N_20 Abatement Project, which is supported by the United Kingdom. This large-scale project has potential to reduce an estimated 473,759 metric tonnes of CO_2 equivalent per annum. Limited understanding of the opportunities associated with CDM projects, high upfront costs for baseline evaluation, and capacity to develop CDM project proposals, among other policy and institutional implementation challenges, have been cited as the main impediments to adoption of CDM interventions in the country [48].

The broad national climate policy regime has also enabled the engendering of climate change in national budgets. **Figure 1** indicates the budgetary allocations given to the CCMD from 2016 to 2019. There has been a significant increase in funds allocated to support the climate change coordination activities of the CCMD since 2016. This increased attention to climate change can be attributed largely to the recognition of the climatic challenge in the recent national economic blueprints, in particular, the Zimbabwe Agenda for Sustainable Socio-Economic Transformation (ZimASSET) (2013–2018) and the Transitional Stabilisation Programme (TSP) (2018–2020). While ZimASSET did not articulate clear guidance on climate change interventions, there has been increased consideration and guidance on climate change in the TSP.

The sustainability of these policy interventions deserves critical analysis. The next section uses the sustainability lens in examining the challenges and opportunities associated with sectoral climate change interventions in Zimbabwe. The policy

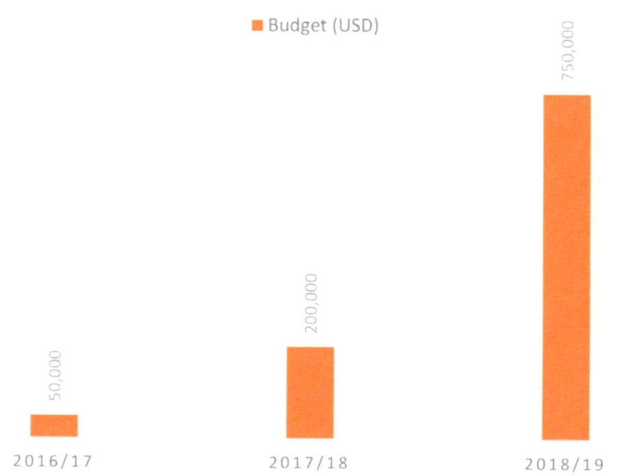

Figure 1.
Treasury allocations to the Climate Change Management Department, 2016–2019 (source: CCMD Official).

environment and the capacity of multi-sectoral institutions responsible for addressing climate change is also discussed.

5. Sectoral interventions, challenges and opportunities

Acknowledging the effects of climate change and guided by the policy and institutional framework presented above, the country, institutions and individuals have not been complacent. The responses, depending on the nature of the climatic event and the persons involved, have either been well-planned or spontaneously executed. This section discusses the sectoral responses to climate change by various stakeholders. Aside from sectoral analysis of the mitigation and adaptation practices in place, the discussion attempts, where possible, to disaggregate the analysis at the level of government, institutions and individuals. The interventions discussed here are intended primarily to identify opportunities that have been unlocked or that are potentially available within the sectors in question. It also ingrains the sustainability question in the analysis.

5.1 Climate change responses in the agriculture sector

There is growing evidence that farmers in Zimbabwe are adapting to observed climate changes. This is through altering cultivation and sowing times and crop cultivars and species that can withstand climatic irregularities. Notable progress in the agriculture sector relates to the development of irrigation infrastructure. In 2001, about 152,000 hectares of land were under formal irrigation with a total of 5000–20,000 under informal irrigation. A further 600,000 ha of land nationwide was to be availed for irrigation development. In 2015, government availed a total of US$2.6 million towards completion of 13 irrigation schemes covering about 635 ha [21]. Since then, government with support from development partners has upscaled the irrigation programme as illustrated in **Table 2**. It is clear from the table that the adaptation agenda has got support largely from international players who have injected funds and equipment to increase the area that can be put under irrigation. The objective is to depart from a practice that has largely relied on rain-fed agriculture to subsequently harness the available water resources for irrigated farming. Within the irrigation policy drive, a number of projects on resilience capacity building in agriculture for communities to better cope with the negative impact of climate change have been implemented. These interventions have also strengthened the Agriculture Extension Services Department (Agritex) to be able to give advisory warnings on planting, crop maturing varieties, including varying of planting dates to spread risks.

Climate change has also led to innovative ways of adaptation in the agriculture sector. These range from isolated practices such as moisture conservation practices by farmers to well-developed responses of CSA such as precise fertiliser application, manure application, agroforestry, crop rotations and intercropping and soil conservation [49]. Adoption of moisture conservation farming practices for example, enable farmers to extend the growing season and to do dual season cropping. In places such as Muzarabani, where climate change has increased the frequency and severity of floods, the practice of dual season cropping has been observed. This strategy enables the locals to harness opportunities associated with flooding [50]. However, floods bring mixed fortunes to the communities experiencing them such as improved soil fertility and ground water recharge [10, 51], but adaptation interventions being practiced in such areas may not be sustainable.

Climate-smart agriculture is farming that embraces the twin goals of mitigation and adaptation at the farm level. The Food and Agriculture Organisation (FAO) describes it as a sustainable climate sensitive response in the agriculture sector with

Responsible authority	Funds (US$)	Target (ha)	Description
European Union (EU)	7.8 million	1206	Technical support for 20 irrigation schemes in Chimanimani, Makoni, Chipinge, Beitbridge, Gwanda and Mangwe districts
Swiss Development Cooperation	1.3 million	656	Rehabilitation of 14 irrigation schemes, benefiting 1425 households in Bikita, Gutu, Masvingo and Zaka districts
Japanese International Cooperation Agency (JICA)	15 million	674	Rehabilitation and development of Nyakomba irrigation scheme in Nyanga District along Gairezi River
International Fund for Agricultural Development (IFAD)	60 million	—	Smallholder Irrigation Revitalisation Programme commencing in 2017
Kuwait and Abu Dhabi funds supported projects	28.7 million	11,290	Maintenance of irrigation schemes for over 2000 households' beneficiaries across the country. This was to be complemented by US$8,6 million of support from development partners. Co-financing of Zhove Irrigation Scheme with government contributing US$7 million
Department for International Development (DFID)	48 million	—	Rehabilitating irrigation schemes for smallholder farmers and supporting training and extension services

Table 2.
Irrigation support projects in Zimbabwe (source: [21]).

co-benefits of increasing productivity and building the resilience of agricultural-based livelihoods communities while reducing GHG emissions. It is a planned intervention strategy encompassing agricultural practices, policies, institutions and financing to bring tangible benefits particularly to smallholder farmers and to enable them to be stewards of the environment that support them [49, 52]. With support from development partners, Zimbabwe has started implementing the CSA programme. The programme targets small-scale farmers, particularly women and poor households that are vulnerable to food insecurity under a changing climate. Elsewhere, successful results have been noted in Kenya and Tanzania as detailed in **Box 1**.

The CSA pilot projects (2011–2014), implemented jointly with partners in Kenya and Tanzania, promoted integrated and diversified farming systems and agro-ecological principles. The programme was established to demonstrate that ongoing agricultural development programmes could bring co-benefits in terms of climate change adaptation and mitigation thereby increase the uptake of CSA at significantly larger scale. The pilot projects linked research activities, practical work in farmers' fields and policy making at different levels to enhance the effectiveness of planning and programming for CSA on farms, throughout the landscape and at the national level. Results showed that: The main benefits of following the CSA approach resulted in higher yields, raised farm income and increased food availability. This is an indication that CSA can be an effective approach for improving food security, alleviating poverty and building more resilient livelihoods. It also indicates that smallholder farmers can be an effective part of the response to climate change and make a meaningful contribution to reducing GHG emissions. Scenarios, modelling and measurements serve an important role in evaluating and prioritising CSA practices for implementation and scaling up. By building research into ongoing development activities, the assessment of CSA practices can be undertaken more quickly, and the findings can be used to prioritise efforts in projects and programmes. Bringing sound, up-to-date evidence into decision-making processes can help shape policy making that effectively supports CSA. The findings from the pilot activities were presented in national workshops, which allowed decision makers to become familiar with the benefits of CSA practices and develop or adjust policies, plans and programmes to better foster CSA.

Box 1.
Successful climate smart agriculture practices in Kenya and Tanzania (source: [49]: xii-xiii).

It is clear from the cited Kenyan and Tanzanian cases that for climatic interventions to be successful, they need to be driven by evidence-based policy formulation and trialled in participatory learning experiences with the concerned communities. Similar CSA approaches, though existing at small isolated scales, are also practiced in Zimbabwe. Many studies carried out in Zimbabwe identify the development of irrigation facilities, growing of small grains and short to medium term crops which mature early and are drought tolerant, and introduction of new agricultural techniques and practices as opportunities farmers were harnessing in adapting to drought [24, 26, 53, 54]. Chanza [53] collaborated with earlier views by Mararike [55] and Kaseke [56] that revival of indigenous food security strategies at village level is an important direction to adapt to climate change disturbances that lead to food insecurity.

The major concern, however, is that most of the climatic support to the farmers has largely been driven by the donor community and the direct support by government is insufficient to meet the assortment of farmers described earlier. As such, farmers who take long to be independent may not be able to continue with the new agricultural techniques without external support.

5.2 Climate change responses in the water sector

The unpredictable and potentially devastating effects of climate change puts a strain on the management of water resources. Zimbabwe's water sector faces mixed challenges such as satisfying increasing competing and conflicting uses owing to climate change effects and increased water demand by other sectors and underutilisation of water resources in some areas. Degradation of water quality worsens the urban water supply situation in the country. This also creates potential for conflict among the different sectors and water users. With proper decisions however, climate change can guide society and water users to be water sensitive and adopt water conservation practices. The challenges related to unpredictable rainfall patterns have seen government, with support from development agencies, investing in irrigation development and maximising on use of existing water and irrigation facilities. Despite the capacity to irrigate more than 330,000 ha, only 80,000 ha were under irrigation in 2016 [21]. There are many ways in which investment in irrigation can bring benefits to the country and farmers involved. For instance, irrigation enables expansion of agriculture activities by turning dry areas into highly productive lands. Development of irrigation infrastructure allows continuous crop production and can facilitate increased productivity where farmers supplement rain fed agriculture. However, under the changing climate, irrigation cannot be business as usual since it is also likely to be affected by the increasing frequency of droughts. The government has moved in to promote centre pivot irrigation to save the water resources and address the high costs associated with the more efficient drip irrigation [57].

Beginning in 2016, Government of Zimbabwe started implementing the Climate Resilient National Water Resources and Irrigation Master Plan, whose objective is to integrate climate change modelling with development and management of water resources and irrigation infrastructure. Under this scheme, the government secured a US$98 million loan facility to buy irrigation equipment, tractors and implements through Brazil's More Food for Africa programme. The programme has been extended to cover small-scale farmers. For instance, following acknowledgement that the available water bodies are being under-utilised, government mooted an integrated water use master plan beginning with Tokwe Mukosi Dam. The plan is expected to support irrigation farming, fisheries, hydropower supply and tourism. The dam reported as the largest inland reservoir in the country, has capacity to irrigate 25,000 ha and can supply 15 MW of hydropower. Clearly, this intervention has managed to resuscitate idle irrigation infrastructure to increase food production. There are also opportunities for technology, knowledge and skills transfer. For example, through using drip and canal irrigation that use less power as compared to the overhead sprinkler methods [21].

With reference to urban areas, the threats of water scarcity associated with climate change have caused water institutions to embark on water saving practices and recycling. It is a fact that urbanisation, whether with or without climate change, imposes increased water use and consumption demands. Accordingly, through adapting water sensitive practices such as recycling, more water can be availed into the supply system. If treated to meet specific water quality standards, wastewater can still be discharged back into public river systems for ecological support and use by downstream communities [10].

The key challenge, however, is that investment in the water sector or in setting up irrigation infrastructure requires large funding. Given the predominantly external based support in irrigation projects that is presented earlier in **Table 2**, there are notable deficiencies in upscaling climatic responses in the water and agriculture sectors. Therefore, unless government allocates adequate funding for irrigation development, the current practice is not only slow in implementation but also not sustainable.

5.3 Climate change responses in the energy sector

The energy sector remains a key intervention focal area by the Government of Zimbabwe. In response to the UNFCCC's global call to cut GHG emissions, Zimbabwe set the conditional mitigation contribution of reducing emissions by 33% below a business as usual (BAU) scenario by 2030. This goal is to be accomplished by uptake of robust responses in the energy sector. Projects that are currently running include ethanol blending, solar water heaters, energy efficiency improvement, increasing hydropower generation in the energy mix, and the refurbishment and electrification of the rail infrastructure. The country is on course to meet these target reductions in carbon-dioxide (CO_2), methane (CH_4) and nitrogen oxide (N_2O) gases. Other mitigation strategies proposed include coal-bed methane power, solar powered off-grids, integrated waste management, changing thermal power station technologies, reviewing the transport system, upscaling the UN-REDD+ implementation and sustainable energy alternatives in the tobacco farming system [46, 58].

As explained earlier, responses in the energy sector are being supported by an enabling policy framework. Specific policies related to the energy sector include the National Climate Policy and the Transport Policy, alongside other climate mitigation instruments. Other policies expected to support GHG mitigation interventions include the Forest Policy, Renewable Energy Policy and Bio-fuels Policy, which are being finalised for adoption. The supportive policy framework has enabled the country to speed up the upgrading of hydro-power generation plants (the recent one being the Kariba Dam Project) and the completion of the Tokwe Mukosi Dam cited earlier. Already the country is on course in renewable energy drive although there are still some challenges to be addressed to scale up the implementation and uptake of renewable energy. Some of these challenges include un-viable tariffs and the low creditworthiness of the power utility who is the major offtaker. **Table 3** shows some of the key projects that are at various stages of implementation, notably the Batoka and the Gairezi hydro-power plants, with others already been completed. The bigger projects capable of generating at least 100 MW have largely been spearheaded by the Zimbabwe Power Company (ZPC), with independent power producers (IPP) concentrating on smaller projects. In addition to the projects indicated in **Table 3**, small hydro-power projects on run off

Project description	Proponent	Energy contribution
Expansion of the Kariba South Power Station	ZPC	300 MW
Batoka Gorge Hydropower Project	Zambezi River Authority (ZRA)	1200 MW (for Zimbabwe) and 1200 MW (for Zambia)
Gwanda Solar Power Plant	ZPC	100 MW
Insukamini Solar Power Plant	ZPC	100 MW
Munyati Solar Power Plant	ZPC	100 MW
Pungwe Hydropower Plant	Nyangani Renewable Energy (IPP)	3 MW
Kupinga Hydropower	IPP	1.4 MW
Gairezi Hydropower Project	ZPC	30 MW

Table 3.
Clean energy project interventions (source: [21]).

river in the Eastern Highlands, and on inland dams around the country are variously taking course [21, 22].

Although still lacking the appropriate supporting policy instruments, fuel blending of E10, E15 and E85 have been introduced. The major challenge is related to limited awareness and low uptake of these products by the public. Solar energy technologies are widely being adopted especially for lighting, powering phones and solar-powered geysers in some households. Most of the large urban areas such as Harare, Bulawayo and Gweru have embarked on projects to use solar-powered traffic lights in the cities although these maybe low-key initiatives compared to what countries like South Africa, Kenya and Morocco have done in the solar energy space. The Rural Electrification Agency (REA) of Zimbabwe has scaled the uptake of solar systems in schools, clinics and public facilities. REA has also supported the uptake of biogas digesters to provide alternative energy for cooking for rural households. Overall, the mitigation initiatives highlighted here present enormous opportunities for a developing country like Zimbabwe. A number of development windows have been opened for international collaboration towards low carbon development pathways and economic development. Investments in low emissions development (LED) are still limited but have potential to grow. Therefore, the country is set to fully benefit from a LED trajectory [22].

In order to respond to the twin problems of energy poverty and land degradation, the Zimbabwe government implemented energy sector reforms that aimed at substituting biomass fuels with liquefied petroleum gas (LPG). Davidson et al. [59] reported a reduction in charcoal use, in favour of LPG consumption, which grew by an annual rate of 12%. The use of LPG also stopped the production of an estimated 337,500 tonnes of charcoal that would have destroyed about 40,500 ha of forest [60]. As argued by Johnson and Lambe [61], switching from a traditional biomass fuel source, for example, charcoal to an environmentally friendly source (LPG) can often lead to adaptive response mechanisms.

A renewable energy project supported by Oxfam and Practical Action in rural areas of Masvingo and Manicaland provinces has yielded positive benefits to the communities. As detailed in **Box 2**, the project has literally energised the beneficiaries as it led to improved health outcomes, widened access to education, increased agricultural production and boosted business and enterprise, strengthened livelihoods, and enhanced quality of life. Already the intervention has shown possibilities of creating green communities that are independent of the national grid and becoming self-sustaining [33].

> The Rural Sustainable Energy Development Project (RuSED) in Zimbabwe ran from August 2011 to January 2016. The project was funded by a two million euros grant from the European Union and Oxfam and was led and implemented by Oxfam in partnership with Practical Action and in association with the Ministry of Energy and Power Development and the Rural Electrification Authority of Zimbabwe. The project aims to enhance the lives and livelihoods of poor rural people by harnessing energy from the sun and running water to bring electricity to remote and isolated communities in ways that are affordable and sustainable. Over the course of the project, Oxfam has implemented a solar energy scheme in Gutu District in Masvingo province, and Practical Action a micro-hydro project in Himalaya in Mature District in Manicaland. The Himalaya scheme was commissioned on 8 April 2015. The Gutu scheme has many elements, including a solar pumping extension to the Ruti irrigation scheme which was commissioned on 10 April 2015.
>
> Results show that access to affordable and reliable electricity from the sun or from running water is crucial to boosting enterprise and increasing production. This has improved quality of life of the beneficiaries, in particular, the quality of women's lives. Access to energy and water has also improved the social and psychological health of communities and their sense of empowerment.

Box 2.
Case study of a clean energy project in Zimbabwe (source: [33]).

The project cited above (see **Box 2**) presents numerous development opportunities for rural development in Zimbabwe. This is a clear demonstration that decentralised energy systems have a potential to contribute to a sustainable future in Zimbabwe. The sustainability of the project has been guaranteed since it enabled communities to take ownership, set their own priorities for energy use and devise payment systems such that they will be able to finance the ongoing operation and maintenance, and ultimately expansion and improvement. Notwithstanding the encouraging progress, much remains to be done in terms of activities to complement energy access that will enable enterprises to thrive [33]. The main challenge could be related to the fall in general economic development indicators that would make it difficult for poor households to access the energy resource.

5.4 Climate change responses in the health sector

Zimbabwe's commitment in the health sector is generally reflected through international, regional and national frameworks. Within these instruments, health issues associated with climate or weather-related shocks and stresses are addressed. Some of the international obligations have been domesticated into national policies and legislation, starting with its Constitution, medium term policies and sectoral strategies in the health sector and in relation to climate change. The NCP and NCCRS give specific mention of health, while the 2016–2020 National Health Strategy makes explicit reference to the need to improve climate change awareness and the need to develop a Public Health Adaptation to Climate Change Plan [22].

Through a strong epidemiological surveillance system in place, the country is capable of giving an early detection of changes in incidence, mortality and geographic range of health outcomes associated with climatic change. One of the critical national programmes to respond to the observed and anticipated spatial spread of malaria mosquitoes is the National Malaria Control Programme, spearheaded by the Department of Disease Prevention and Control in the Ministry of Health and Child Care. The programme implements many strategies, including vector control, case management, epidemic preparedness and response, intermittent preventive therapy, research, monitoring and evaluation, and information, education and advocacy for malaria treatment and prevention. The programme receives support from two major donors: The Global Fund to Fight AIDS, Tuberculosis and Malaria and the President's Malaria Initiative [22].

In some places such as Muzarabani, it can be argued that the desiccation of wetlands and ponds that previously harboured vectors and acted as breeding grounds for mosquitoes has significantly reduced disease incidences. Drought has also led to serious water scarcities prompting the government and other development partners to sink boreholes in order to improve access to portable water. This means people can now easily access portable water, which previously they could not. In this thinking therefore, climate change is arguably an opportunity for community development through interventions to improve water and sanitation [50]. The major threat to this drive emanates largely from the depletion of ground water sources described earlier. This means communities in some dry regions may only have seasonal access to the portable water as drought events worsen.

Existing policies also create adaptation opportunities that can assist in evading adaptation barriers. Worth mentioning is the National Water Policy of 2012 that provides an enabling environment for climate change response. Within it, the Zimbabwe National Water Act specifies the need to use water efficiently and applies a user pays principle that regulates water use. Alongside other development policies, the water policy aims to promote uptake of cleaner and more efficient technologies across all water consumption sectors. This has seen sectoral and institutional collaborations in funding the construction of solar powered boreholes in dry areas of the country such as Chivi District in Masvingo Province. This has been supported by construction of Blair toilets to improve the hygiene and sanitation of the communities [22]. Zimbabwe's Water, Sanitation and Hygiene (WASH) sector is managed and coordinated by an inter-ministerial committee, the National Action Committee (NAC) with the National Coordinating Unit (NCU) as the Secretariat. WASH components comprising of Hygiene Promotion, Water Supply, Excreta Disposal, Vector Control, Solid Waste Management and Drainage require protection from damage and disruption by climate change induced disasters. Should they be damaged, they urgently require restoration to avert deaths, diseases and malnutrition. The NAC has been strategic in engaging partners, mobilising resources and ensuring timeous response to WASH disasters. Undoubtedly, the sector interventions lessen the impacts of climate change as people have access to adequate water supply and sanitary facilities, which are key provisions in reducing diarrhoea and other infectious diseases.

6. Discussion

Emerging from this exposition is that if exploited well the potential benefits of climate change could be realised in all the socio-economic development sectors discussed in this article. The country needs to identify the best alternatives that do not involve lots of capital and are adaptive to local communities in Zimbabwe. With reference to the agriculture sector, there are opportunities for livelihoods diversification pointed out by Chikodzi et al. [24] and Chanza [53] where adaptation on ensuring food security under climate change could have the most direct benefits on livelihoods. There are also multiple benefits for food security, including enhancing food production, access to markets and resources, and reduced disaster risk. Effective adaptation of cropping can help ensure food production and thereby contribute to food security and sustainable livelihoods by enhancing current climate risk management. It is also important to point out that climate change has allowed climate sensitive budgeting in the broad socio-economic development sectors of the country.

The situation in benefits of climate adaptation in the water sector appears blurred. There are places, particularly in urban areas, which are expected to experience serious water supply challenges while other areas, mainly rural communities, are evidently harnessing opportunities brought about by climatic events.

From a social development perspective, water and sanitation interventions have also impacted on the gender dimension of the rural community. In a study in Muzarabani, one of the dryland rural community largely regarded as the epicentre of climatic disturbances [51, 62, 63], women and girls who used to travel long distances to access water are now travelling less distances owing to proximity and improved access of portable water from boreholes drilled in their villages [50].

Effective responses in the energy sector tend to be constrained by limited funding for project development, lack of feasibility studies for wind power generation to prove the achievable capacity, lack of financing to upgrade feasibility studies of some small hydropower sites which were carried out back in the 1990s and lake of capacity to install and maintain renewable energy systems. In addition, there are weaknesses in institutional capacity for support mechanisms. Notwithstanding the existing challenges, there are promising nuances in unlocking development benefits to the country, institutions and individuals. Specifically, Zimbabwe should strengthen the policy on energy efficiency along with supporting instruments that can be used to support energy efficiency adoption by industries. The country is challenged to fully develop the market for energy services. Apparently, decision makers lack in awareness on markets for energy services; service providers are unable to deliver the appropriate market services to unlock the market for energy services; and financiers are not appreciative of the energy efficiency business and therefore fail to deliver sustainable financial products. Elsewhere, it has been proven that putting in place mechanisms for accessing energy efficient technologies can create energy security, energy access, employment generation, cost-savings and health benefits to countries adopting such a practice [64, 65].

In order to fully realise the benefits of climate change in the health sector, the government would need to strengthen its health warning systems on climate-related disturbances. Generally, the early warning systems (EWS) are still weak as they tend to be poorly supported by early action. In other places, the health sector has employed EWS to predict disease for adaptation planning and implementation [66]. For example, studies done in some parts of Africa have assisted in predicting conditions expected to lead to an outbreak of Rift Valley fever [67] and in predicting meningitis against weather and climatic extremes [68] to facilitate early disease interventions. Confalonieri et al. [66] indicate that through public awareness, individual-level responses and adaptation to climate change can be improved. It has been established that the effectiveness of health warning systems, for extreme events such as heat waves and floods, depends on individuals taking appropriate actions [66, 69]. Hence, to achieve maximum benefit from climate response, it is imperative that the disaster affected population has the necessary information, knowledge and understanding to take appropriate action.

Health benefits of responding to climate change are well documented. Ludi et al. [17] explain that health benefits can be achieved by greener and more sustainable choices in broad sectors covering household energy, electricity generation, transport, urban planning and land use, buildings, food and agriculture. For example, the use of cleaner fuels and cooking technologies can reduce the large burden of disease from household air pollution in developing countries; greater use of renewables in electricity generation can cut ambient air pollution; behavioural shifts towards walking and cycling can reduce the burdens of both physical inactivity and air pollution [70–72]. Zimbabwe can take advantage of existing health sector interventions such as public education and awareness campaigns to reduce the risk of diarrhoeal and vector-borne diseases whose incidences may be worsened by climate change. Accordingly, adaptation strategies to climate change in the health sector can result in development of capacity building to evade barriers associated with climate change. If attention can be given to such critical institutions as the Meteorological Services Department (MSD), rural district councils (RDCs) and health institutions, for example through funding, the health system can be strengthened to address climatic challenges.

Overall, although a range of benefits are evident in the broad socio-economic development sectors of the country, much still needs to be done to enhance the sustainability trajectory of climate change responses. Most of the interventions discussed here have mainly been spearheaded by the external driven initiatives, mainly in terms of policy direction and funding. The main reasons for an external driven orientation relate largely to the macro-economic problems that the country has been facing over the past two decades and partly to the heterogeneous acknowledgement of climate change as a development priority on the policy and institutional front.

7. Conclusion

From the analysis given in this article, Zimbabwe, like many developing countries, faces climate change in its main socio-economic development sectors. Although the article only concentrated on the agriculture, water, energy and health sectors to show climate change impacts and the country's responses to the climate agenda, it is proper to conclude that climate change brings mixed experiences that need to be carefully studied. The study challenges the current discourse that have tended to project climate change as a development hindrance. Instead, the article revealed several development opportunities that exist. If these opportunities are carefully considered, government, communities and individuals will be able to take advantage of the climate change phenomena to reshape the development trajectory. At the policy front, climate change has intensified policy formulation whose benefits go beyond environmental to cover co-benefits in the broad socio-economic development sectors. This has unlocked investment opportunities in clean energy and the associated health benefits, improved energy access, improved energy security particularly in remote and newly developed settlements, access to portable water, expansion of irrigation facilities, climate-sensitive budgeting, improved agriculture production, and improved food security. Essentially, most of the climatic interventions associated with these benefits have also managed to articulate cross-cutting issues of gender, poverty and marginalised groups. In the context of the energy sector, communities that are otherwise far from the grid could benefit as they get closely connected to the world through off-grid energy systems, modern communications and information technology. Clearly, the climate response regime that Zimbabwe embraces has opened up several avenues for addressing poverty. However, these benefits are not evenly experienced but tend to be isolated across the Zimbabwean communities.

The sustainability question on whether the current and anticipated benefits of climate change responses can be guaranteed to continue accruing to individuals, institutions and the country at large has been investigated. One way of making climatic responses sustainable would be to leverage the current predominantly external funding to get the necessary knowledge and best practices implemented to inform necessary government budgetary allocations that is supported by climate-sensitive development plans and policies. The government should depart from external funding but promote blended financing approach to allow for ownership and enhance impact investment by all players. Essentially, benefits of responding to climate change are only fully realised when the country embraces both mitigation and adaptation practices in its response decision mix. Mitigation should not only be understood as concerned with cutting carbon emissions but should be designed to take advantage of technological advances in renewable energy, for example, among other opportunities that it offers.

Author details

Nelson Chanza[1]* and Veronica Gundu-Jakarasi[2]

1 Department of Geography, Bindura University of Science Education, Bindura, Zimbabwe

2 Infrastructural Development Bank of Zimbabwe (IDBZ), Harare, Zimbabwe

*Address all correspondence to: nchanza@gmail.com

References

[1] IPCC. In: Field CB, Barros VR, Dokken DJ, Mach KJ, Mastrandrea MD, Bilir TE, editors. Climate change 2014: Impacts, adaptation, and vulnerability. Part A: Global and sectoral aspects, Contribution of Working Group II to the Fifth Assessment Report of the Intergovernmental Panel on Climate Change. Cambridge and New York: Cambridge University Press; 2014. 1132 p

[2] Hallegatte S, Bangalore M, Bonzanigo L, Fay M, Kane T, Narloch U, et al. Shock waves: Managing the impacts of climate change on poverty. In: Climate Change and Development Series. Washington, DC: World Bank; 2006. DOI: 10.1596/978-1-4648-0673-5

[3] Serdeczny O, Adams S, Baarsch F, Coumou D, Robinson A, Hare W, et al. Climate change impacts in sub-Saharan Africa: From physical changes to their social repercussions. Regional Environmental Change. 2017;**17**:1585. DOI: 10.1007/s10113-015-0910-2

[4] Escarcha JF, Lassa JA, Zander KK. Livestock under climate change: A systematic review of impacts and adaptation. Climate. 2018;**6**:54. DOI: 10.3390/cli6030054

[5] Smith MR, Myers SS. Impact of anthropogenic CO_2 emissions on global human nutrition. Nature Climate Change. 2018;**8**:834-839. DOI: 10.1038/s41558-018-0253-3

[6] IPCC. Summary for policymakers. Climate change 2014: Mitigation of climate change. In: Edenhofer O, Pichs-Madruga R, Sokona Y, Farahani E, Kadner S, Seyboth K, Adler A, Baum I, Brunner S, Eickemeier P, Kriemann B, Savolainen J, Schlomer S, von Stechow C, Zwickel T, Minx JC, editors. Contribution of Working Group III to the Fifth Assessment Report of the Intergovernmental Panel on Climate Change. Cambridge and New York: Cambridge University Press; 2014

[7] FAO. Regional Overview of Food Insecurity: African Food Security Prospects Brighter than Ever. Accra: FAO; 2015

[8] Brazier A. Climate Change in Zimbabwe: Facts for Decision Makers and Planners. Konrad-Adenauer-Stiftung: Harare; 2015

[9] Government of Zimbabwe (GoZ). Zimbabwe National Climate Change Response Strategy. Harare. Water and Climate: Ministry of Environment; 2015

[10] Chanza N. Limits to climate change adaptation in Zimbabwe: Insights, experiences and lessons. In: Filho WL, Nalau J, editors. Limits to Adaptation: Insights and Experiences. Climate Change Management Series. Cham. Springer; 2018. pp. 109-127. DOI: 10.1007/978-3-319-64599-5_6

[11] Wilbanks TJ, Sathaye J. Integrating mitigation and adaptation as responses to climate change: A synthesis. Mitigation and Adaptation Strategies for Global Change. 2007;**12**:957-962. DOI: 10.1007/s11027-007-9108-3

[12] Tol RSJ. Adaptation and mitigation: Trade-offs in substance and methods. Environmental Science & Policy. 2005;**8**:572-578. DOI: 10.1016/j.envsci.2005.06.011

[13] UNFCCC. Paris Agreement. United Nations Framework Convention on Climate Change [Internet]. 2015. Available from: https://unfccc.int/sites/default/files/english_paris_agreement.pdf [Accessed: 15 October 2018]

[14] Boko M, Niang I, Nyong A, Vogel C, Githeko A, Medany M, et al. Africa. In: Parry ML, Canziani OF, Palutikof JP,

Van der Linden PJ, Hanson CE, editors. Climate Change 2007: Impacts, Adaptation and Vulnerability. Contribution of Working Group II to the Fourth Assessment Report of the Intergovernmental Panel on Climate Change. Cambridge: Cambridge University Press; 2007. pp. 433-467

[15] Niang IOC, Ruppel MA, Abdrabo A, Essel C, Padgham LJ, Urquhart P. Africa. In: Barros VR, Field CB, Dokken DJ, Mastrandrea MD, Mach KJ, Bilir TE, Chatterjee M, Ebi KL, Estrada YO, Genova RC, Girma B, Kissel ES, Levy AN, MacCracken S, Mastrandrea PR, White LL, editors. Climate Change 2014: Impacts, Adaptation, and Vulnerability. Part B: Regional Aspects. Contribution of Working Group II to the Fifth Assessment Report of the Intergovernmental Panel on Climate Change. Cambridge and New York: Cambridge University Press; 2014. pp. 1199-1265

[16] Klein RJT, Midgley GF, Preston BL, Alam M, Berkhout FGH, Dow K, et al. Adaptation opportunities, constraints, and limits. In: Field CB, Barros VR, Dokken DJ, Mach KJ, Mastrandrea MD, Bilir TE, Chatterjee M, Ebi KL, Estrada YO, Genova RC, Girma B, Kissel ES, Levy AN, MacCracken S, Mastrandrea PR, White LL, editors. Climate Change 2014: Impacts, Adaptation, and Vulnerability. Part A: Global and Sectoral Aspects. Contribution of Working Group II to the Fifth Assessment Report of the Intergovernmental Panel on Climate Change. Cambridge and New York: Cambridge University Press; 2014. pp. 899-943

[17] Ludi E, Jones L, Levine S. Changing Focus? How to Take Adaptive Capacity Seriously. Evidence from Africa Shows that Development Interventions Could Do More. ODI Briefing Paper 71. London: Overseas Development Institute (ODI); 2012. 4 p

[18] Adger WN. Social capital, collective action, and adaptation to climate change. In: Voss M, editor. Der Klimawandel. VS Verlag für Sozialwissenschaften. 2010

[19] Mtisi S, Prowse M. Baseline Report on Climate Change and Development in Zimbabwe. Government of Zimbabwe: Harare; 2012

[20] UN (United Nations). Sendai Framework for Disaster Risk Reduction 2015-2030. Geneva [Internet]. 2015. Available from: https://www.unisdr.org/files/43291_sendaiframeworkfordrren.pdf [Accessed: 15 October 2018]

[21] Government of Zimbabwe (GoZ). The 2016 National Budget Statement. Harare: Ministry of Finance and Economic Development. 2016

[22] Government of Zimbabwe/United Nations Development Programme (GoZ/UNDP). Zimbabwe Human Development Report: Climate Change and Human Development: Towards Building a Climate Resilient Nation [Internet]. 2017. Available from: hdr.undp.org/sites/default/files/reports/2842/undp_zw_2017zhdr_full.pdf [Accessed: 20 October 2018]

[23] Mugandani R, Wuta M, Makarau A, Chipindu B. Re-classification of agro-ecological regions of Zimbabwe in conformity with climate variability and change. African Crop Science Journal. 2012;**20**(s2):361-369

[24] Chikodzi D, Murwendo T, Simba FM. Climate change and variability in Southeast Zimbabwe: Scenarios and societal opportunities. American Journal of Climate Change. 2013;**2**:36-46

[25] Mutasa C. Evidence of climate change in Zimbabwe. In: Proceedings of the Climate Change Awareness and Dialogue Workshop for Mashonaland Central and Mashonaland West

Provinces; 29-30 September 2008; Kariba: Caribbea Bay Hotel

[26] Unganai LS. Adaptation to climate change among agropastoral systems: Case for Zimbabwe. Earth and Environmental Science. 2009;**6**(41):412045. DOI: 10.1088/1755-1307/6/1/412045

[27] Jiri O, Mafongoya PL, Chivenge P. Contextual vulnerability of rainfed crop-based farming communities in semi-arid Zimbabwe: A case of Chiredzi District. International Journal of Climate Change Strategies and Management. 2017;**9**(6):777-789. DOI: 10.1108/IJCCSM-03-2017-0070

[28] Tubiello FN, Soussana JF, Howden SM. Crop and pasture response to climate change. PNAS. 2007;**104**(50):19686-19690

[29] Davis R, Hirji R. Climate Change and Water Resources Planning, Development and Management in Zimbabwe. An Issues Paper. World Bank; 2014

[30] Kjeldsen TR, Lundorf A, Rosbjerg D. Use of a two-component exponential distribution in partial duration modelling of hydrological droughts in Zimbabwean rivers. Hydrological Sciences Journal. 2000;**45**(2):285-298. DOI: 10.1080/02626660009492325

[31] IPCC. Analysing regional aspects of climate change and water resources: Africa [internet]. 2008. Available from: www.ipcc.ch/pdf/technical-papers/ccw/article5.pdf [Accessed: 10 October 2018]

[32] IPCC. In: Edenhofer O, Pichs-Madruga R, Sokona Y, Seyboth K, Matschoss P, Kadner S, Zwickel T, Eickemeier P, Hansen G, Schlömer S, von Stechow C, editors. Summary for Policymakers. IPCC Special Report on Renewable Energy Sources and Climate Change Mitigation. Cambridge and New York: Cambridge University Press; 2011

[33] Magrath J. Transforming Lives in Zimbabwe: Rural Sustainable Energy Development Project. Oxfam, GB [internet]. 2015. Available from: https://www.scribd.com/document/340882047/Transforming-Lives-in-Zimbabwe-Rural-Sustainable-Energy-Development-Project [Accessed: 12 September 2018]

[34] Kusangaya S, Warburton ML, van Garderen EM, Jewitt GPW. Impacts of climate change on water resources in Southern Africa: A review. Physics and Chemistry of the Earth. Parts A/B/C. 2014;**67-69**:47-54. DOI: 10.1016/j.pce.2013.09.014

[35] ESI-Africa. Drought affects hydroelectricity generation in Kenya [internet]. 2018. Available from: www.esi-africa.com/drought-affects-hydroelectricity-kenya/ [Accessed: 12 October 2018]

[36] Government of Zimbabwe (GoZ). Zimbabwe's Third National Communication to the United Nations Framework Convention on Climate Change. Harare: Ministry of Environment, Water and Climate; 2016

[37] Karoly DJ. Ozone and climate science. Science. 2003;**302**(5643):236-237. DOI: 10.1126/science.1090851

[38] Manatsa D, Mukwada G. A connection from stratospheric ozone to El Niño-southern oscillation. Nature Scientific Reports. 2017;7:5558. DOI: 10.1038/s41598-017-05111-8

[39] Terry G. No climate justice without gender justice: An overview of the issues. Gender and Development. 2009;**17**(1):5-18. DOI: 10.1080/13552070802696839

[40] Muzenda-Mudavanhu C. A review of children's participation in disaster

risk reduction. Jàmbá: Journal of Disaster Risk Studies. 2016;**8**(1), Art. #218, 6 p. DOI: 10.4102/jamba.v8i1.218

[41] Hartmann J, Ebi K, McConnell J, Chan N, Weyant JP. Climate suitability: For stable malaria transmission in Zimbabwe under different climate change scenarios. Global Change and Human Health. 2002;**3**:42-54

[42] Matawa F, Murwira KS. Predicting future habitats of disease vectors using climate change models: The case of the tsetse fly (*Glossina* spp.) in the Matusadona area, North-western Zimbabwe. In: Proceedings of the First Climate Change Symposium of Zimbabwe; 17-19 June 2013; Harare: Cresta Lodge

[43] Olu O, Charimari L, Manangazira P, Ameda I, Shambare D, Midzi S, et al. Unprecedented outbreak of cholera in Zimbabwe: Implications and lessons learned for cholera prevention and control in chronic humanitarian crises [internet]. 2011. Available from: www.uonbi.ac.ke/journals/files/journals/1/.../1025-3764-1-RV.doc [Accessed: 12 October 2018]

[44] WHO. Cholera—Zimbabwe [Internet]. 2018. Available from: http://www.who.int/csr/don/05-october-2018-cholera-zimbabwe/en/ [Accessed: 12 October 2018]

[45] Ministry of Health and Child Care (MHCC). Manicaland Cyclone Idai Health Sector Situation Report, Issue 15, Harare. 2019

[46] UNFCCC. Zimbabwe's Intended Nationally Determined Contribution (INDC) Submitted to the United Nations Framework Convention on Climate Change (UNFCCC) [Internet]. 2015. Available from: www4.unfccc.int/ndcregistry/PublishedDocuments/.../MOZ_INDC_Final_Version.pdf [Accessed: 15 October 2018]

[47] Chanza N, Mundoga T, Nyamudeza S. Deepening understanding of multilateral environmental agreements (MEAs) for policy intervention by the Environmental Management Board. Harare: Environmental Management Agency; 2018

[48] Zhakata W. Climate Change in Zimbabwe—On the Role of Clean Development Mechanism [Internet]. 2007. Available from: www.ccs-africa.org/fileadmin/ccs-africa/user/docs/Gabarone_10_9/Gaborone_Zhakata_10sept07panel.pdf [Accessed: 25 September 2018]

[49] Nyasimi M, Amwata D, Hove L, Kinyangi J, Wamukoya G. Evidence of Impact: Climate-Smart Agriculture in Africa. CCAFS Working Paper no. 86. Copenhagen, Denmark: CGIAR Research Program on Climate Change, Agriculture and Food Security (CCAFS) [Internet]. 2014. Available from: cgspace.cgiar.org/handle/10568/51374 [Accessed: 12 September 2018]

[50] Chanza N. Indigenous knowledge and climate change: insights from Muzarabani, Zimbabwe [thesis]. Port Elizabeth: Nelson Mandela University; 2014

[51] Gwimbi P. Linking rural community livelihoods to resilience building in flood risk reduction in Zimbabwe. Journal of Disaster Risk Studies. 2009;**2**(1):71-79

[52] FAO. Planning, implementing and evaluating Climate-smart agriculture in smallholder farming systems: The experience of the MICCA pilot projects in Kenya and the United Republic of Tanzania. In: Mitigation of Climate Change in Agriculture Series. FAO: Rome; 2016

[53] Chanza N. Indigenous-based adaptation: An imperative for sustainable climate change strategies for Africa. In:

Mawere M, Awuah-Nyamekye S, editors. Harnessing Cultural Capital for Sustainability: A Pan Africanist Perspective. Bamenda: Langaa Publishing House; 2015. pp. 85-134

[54] Jiri O. Climate change and variability impacts on crop production in the low potential smallholder farming regions of Zimbabwe [thesis]. Pietermaritzburg: University of KwaZulu-Natal; 2016

[55] Mararike CG. Revival of indigenous food security strategies at the village level: The human factor implications. Zambezia. 2001;**XXVIII**(i):53-65

[56] Kaseke E. The revival of Zunde raMambo in Zimbabwe. Focus. 2006;**2**(1). Available from: http://www.vosesa.org.za/focus/vol1_no4/index.html?article_3.html~content [Accessed: 10 September 2018]

[57] Sunday News. Govt Works on $60m Irrigation Facility; 20 May 2019; Harare. 2019

[58] Chanza N, Chirisa I, Zhakata W, Makura ESMS. Ethical and justice reflections in Zimbabwe's INDC and climate policies. In: Brown DA, Taylor P, editors. Lessons Learned from Research on How 23 Nations Actually Considered or Ignored Ethics and Justice in Formulating National Climate Commitments [Internet]. 2015. Available from: ethicsandclimate.org/category/indcs/ [Accessed: 20 September 2018]

[59] Davidson O, Halsnæs K, Huq S, Kok M, Metz B, Sokona Y, et al. The development and climate nexus: The case of Sub-Saharan Africa. Climate Policy. 2003;**3**:S97-S113. DOI: 10.1016/j.clipol.2003.10.007

[60] UNDP. Energizing the Millennium Developing Goals: A Guide to Energy's Role in Reducing Poverty. New York, NY: UNDP; 2005

[61] Johnson FX, Lambe F. Energy Access, Climate and Development. Stockholm Environment Institute, Commission on Climate Change and Development [internet]; 2009. Available from: http://environmentportal.in/files/ccd_energyaccessclimateanddev2009.pdf [Accessed: 12 October 2018]

[62] Mudavanhu C, Bongo PP. Children's coping with natural disasters: Lessons from floods and droughts in Muzarabani District, Zimbabwe. Children, Youth and Environments. 2015;**25**(3):196-2013. DOI: 10.7721/chilyoutenvi.25.3.0196

[63] Mavhura E, Manyena B, Collins AE. An approach for measuring social vulnerability in context: The case of flood hazards in Muzarabani district, Zimbabwe. Geoforum. 2017;**86**:103-117

[64] Zhang Y, Shi H. From burden-sharing to opportunity-sharing: Unlocking the climate negotiations. Climate Policy. 2014;**14**(1):63-81. DOI: 10.1080/14693

[65] Tawney L, Miller M, Bazilian M. Innovation for sustainable energy from a pro-poor perspective. Climate Policy. 2015;**15**(1):146-162. DOI: 10.1080/14693062.2013.781456F

[66] Confalonieri U, Menne B, Akhtar R, Ebi KL, Hauengue M, Kovats RS, et al. Human health. In: Parry ML, Canziani OF, Palutikof JP, Van der Linden PJ, Hanson CE, editors. Climate Change 2007: Impacts, Adaptation and Vulnerability. Contribution of Working Group II to the Fourth Assessment Report of the Intergovernmental Panel on Climate Change. Cambridge. UK: Cambridge University Press; 2007. pp. 391-431

[67] Anyamba A, Linthicum KJ, Small J, Britch SC, Pak E, De La Rocque S, et al. Prediction, assessment of the Rift Valley fever activity in east and southern Africa 2006-2008 and possible vector

control strategies. American Journal of Tropical Medicine and Hygiene. 2010;**83**(2 Suppl):43-51. DOI: 10.4269/ajtmh.2010.09-0289

[68] Cuevas LE, Jeanne I, Molesworth A, Bell M, Savory EC, Connor SJ, et al. Risk mapping and early warning systems for the control of meningitis in Africa. Vaccine. 2007;**25**(Suppl. 1):A12-A17

[69] Grothmann T, Patt A. Adaptive capacity and human cognition: The process of individual adaptation to climate change. Global Environmental Change. 2005;**15**:199-213

[70] Haines A, Kovats RS, Campbell-Lendrumb D, Corvalanb C. Climate change and human health: Impacts, vulnerability and public health. Public Health. 2006;**120**:585-596

[71] Few R. Health and climatic hazards: Framing social research on vulnerability, response and adaptation. Global Environmental Change. 2007;**17**:281-295

[72] Mani M, Climate Change WL. Human Impacts: How Vulnerable Is Bangladesh and What Needs to be Done? Washington, DC: World Bank; 2014

Section 2

Managing Activities

Chapter 4

Mathematical Model for CO_2 Emissions Reduction to Slow and Reverse Global Warming

Nizar Jaoua

Abstract

This chapter aims to provide climate policy makers with smooth patterns of global carbon dioxide (CO_2) emissions consistent with the UN climate targets. An accessible mathematical approach is used to design such models. First, the global warming is quantified with time to determine when the climate targets will be hit in case of no climate mitigation. Then, the remaining budget for CO_2 emissions is derived based on recent data. Considering this for future emissions, first proposed is an exponential model for their rapid reduction and long-term stabilization slightly above zero. Then, suitable interpolations are performed to ensure a smooth and flexible transition to the exponential decline. Compared to UN climate simulation models, the designed smooth pathways would, in the short term, overcome a global lack of no-carbon energy and, in the long term, tolerate low emissions that will almost disappear as soon as desired from the 2040s with no need for direct removal of CO_2.

Keywords: atmospheric carbon dioxide (CO_2), global CO_2 emissions, global warming, remaining CO_2 budget, time model, UN climate target

1. Introduction

The climate change has been declared as an urgent global threat [1] since the spread of devastating floods, severe droughts, and ravaging wildfires, due to rising temperatures especially in the past three decades [2–4]. In response to this threat, the UN parties adopted the *2015 Paris Agreement on Climate Change* along with its implementation by 2020. Was included 'holding the increase in global annual average temperature above the pre-industrial level well below 2°C and pursuing efforts to limit it to 1.5°C' ([1], Art.2). Was also comprised 'projecting global peaking of greenhouse gases emissions as soon as possible along with their rapid reduction' ([1], Art.4).

Since the last century, the atmospheric carbon dioxide (CO_2) has been largely dominating the other greenhouse gases [5, 6] due to increasing anthropogenic CO_2 emissions as a consequence of a growing global demand for fossil-fuel-based products. Subsequently, climate policies would include a massive reduction of these emissions by shifting to no-carbon energy and introducing gas capture/removal technologies.

Climate mathematical modelling has so far focused on the physics behind the global warming, and has therefore described the rise in global average temperature using a mathematical approach based on the law of conservation of energy (e.g., [7–10]). When it comes to the climate mitigation, the existing models were mostly produced by computer simulation, which involved rather climatologists. Among these are the representative concentration pathways (RCPs; [5, 11–13]), which were adopted by the Intergovernmental Panel on Climate Change (IPCC) to predict future annual CO_2 emissions by simulating representative mitigation scenarios of radiative forcing; 2.6, 4.5, 6, and 8.5 W m^{-2}, going from the highest to the lowest mitigation. In the same setting, the C4MIP as part of the CMIP (Coupled Model Inter-comparison Project) provided a set of earth system models, involving the carbon cycle [14], also adopted by the IPCC (AR5, WG I). Among these were included models that infer CO_2 emissions based on atmospheric CO_2 concentrations targets. More recently, mixed models were developed using a combination of simulation climate and socio-economic models [15] to limit the radiative forcing to 1.9 W m^{-2}, and hence to meet the 1.5°C target.

The purpose of this chapter is to provide climate policy makers with smooth patterns of global CO_2 emissions consistent with a prescribed UN climate target, i.e., a limit $\mathcal{L}$ (°C) to the rise in global average temperature above the pre-industrial level. Unlike in literature where modelling is often based on computer simulation, an accessible mathematical analysis is used to design such models. Basically, two parameters are required; an estimation of the emissions level in the beginning of their mitigation (fixed parameter) and the remaining CO_2 budget (dependent parameter) which, by definition, consists of the cumulative CO_2 emissions (from the starting time) that will raise the global average temperature up to the given climate target. These parameters will be determined using a very strong and highly significant linear regression involving recent data on the gas emissions [16], which also provides a time model for these emissions in case of no climate policy. Based on this model, the second parameter will be explicitly determined in terms of the climate target. Modelling future emissions to make them fit the given UN target would be nothing else but connecting their initial state (predicted level in the beginning of the mitigation) to their desired final state (zero or almost-zero emission). Naturally, an exponential interpolation would provide such a connecting way with a rapid reduction of the emissions over the first 50–60 years, their stabilization slightly above zero in the long term, along with their extinction in far future due to the asymptotic behavior of the exponential model. Another source of mathematical modelling with regards to climate mitigation is the transition to this exponential trend, which can provide more feasible patterns for CO_2 emissions. Indeed, an independent parameter is introduced as an arbitrary fraction of the remaining CO_2 budget expected to be used exponentially, which also gives an indication for the transition length. Then suitable quadratic interpolations are performed to smoothly connect the current linear trend to the exponential decline. As a result, an uncountable range of exponential pathways is designed with smooth and flexible transition, which will not only overcome a global shortage of no-carbon energy but also lead to the nearly-zero emission as soon as desired depending on the climate target. The graphical representation of the designed models will help to explore their similarities to the (IPCC) RCPs and no- and low-overshoot 1.5°C pathways [17].

The rest of the chapter is organized as follows. The required materials are presented in Section 2, including recent annual data on CO_2 (concentration in the air as well as emissions level) and the correlation between the global warming and the atmospheric CO_2. In Section 3, a time model for global warming is presented along with a formulation of the hitting time for a given UN climate target. Section 4 is devoted to the elaboration and discussion of smooth mathematical models for global CO_2 emissions consistent with the UN climate targets. The results are summarized in Section 5.

2. Background materials

It is well-known that the global warming is due to the growing concentration of the greenhouse gases in the atmosphere, particularly the anthropogenic CO_2. Its quantification with time would therefore require the consideration of both; its correlation with and annual data of the atmospheric CO_2. On the other hand, recent data on global CO_2 emissions will be necessary to design appropriate pathways for these emissions in order to limit their warming effect to a prescribed UN target. Additionally, non-linear interpolations are inevitable to ensure a smooth transition from the current trend to the rapid decline of the emissions as urged by the UN.

2.1 Global warming vs. atmospheric CO_2

One of the key results in [10], reminded below, will be of great use in modelling with time the global warming. Based on the physics law of conservation of energy, this result states that the rise in global average temperature above the pre-industrial record is growing with the ratio r of CO_2 concentration to the pre-industrial level. It can be seen as a generalization of the well-investigated climate response to doubling CO_2 concentration [5, 18, 19].

$$\Delta T(r) \approx \beta \frac{r-1}{r-k} \qquad (\beta \approx 5.84, k \approx -0.85) \tag{1}$$

2.2 Atmospheric CO_2 data (2000–2017)

The warming effect of the atmospheric CO_2 as quantified in (1), along with the trend of its concentration over the past two decades, will allow to describe the global warming through time. This trend will be estimated by linear regression of the annual average concentration of the gas based on the NASA monthly measurements from 2000 to 2017 [20].

2.3 Global CO_2 emissions data (2000–2013)

It is necessary to determine the trend of the global CO_2 emissions over the past two decades prior to modelling with time the desired effect of any projected mitigation in line with the UN climate goal. This trend will be estimated by linear regression of the annual gas emissions recorded by Carbon Dioxide Information Analysis Center (CDIAC) up to 2013 [16].

2.4 Quadratic interpolation

Classically, a quadratic interpolation consists of determining a quadratic function using the values that it takes on at exactly three particular values of its variable. The following result provides an original quadratic interpolation using also three given data on the parabola representing the function: its symmetry axis, one of its points (other than the vertex), and the slope of the tangent line at that point. This technique will be used to add a smooth transition to an exponential model for CO_2 emissions.

If a parabola is symmetric about the line: $x = u$, *passes through a point* (x_0, y_0), *with* $x_0 \neq u$, *and is tangent at this point to the line of slope* m, *then an equation of this parabola is:*

$$\begin{aligned} y &= A(x-u)^2 + B \\ A &= m/2(x_0-u) \\ B &= y_0 - m(x_0-u)/2 \end{aligned} \tag{2}$$

Indeed, the form of the equation is due to the symmetry about the line $x = u$. The coefficient A is determined by equating the slope m with $\left.\frac{dy}{dx}\right|_{x=x_0} = 2A(x_0 - u)$. Then B is deduced by plugging in x_0, y_0 and A in Eq. (2).

3. Time model for global warming

As it can be seen from (1), an estimation of the atmospheric CO_2 concentration ratio through time will give a time model for the global warming. This estimation can be done by linear regression of the annual average ratio for CO_2, using the NASA dataset [20]. This leads to the following no-climate-mitigation model r_0, applicable from the year 2000 (i.e., $t = 0$)

$$\begin{aligned} &r_0(t) \approx a_0 t + b_0 \\ &a_0 \approx 0.0076, \quad b_0 \approx 1.3176 \end{aligned} \tag{3}$$

Such a linear regression was found to be statistically highly significant ($p < 10^{-21}$) and extremely strong ($r^2 \approx 0.99$). By composing the energy-balance-based model ΔT (given in (1)) with r_0, one gets the following climate-policy-free model w_0 for global warming, applicable from the year 2000 (i.e., $t = 0$)

$$\begin{aligned} \boldsymbol{w_0(t)} &= \boldsymbol{\Delta T(r_0(t)) \approx \beta \frac{t + \lambda_0}{t + \mu_0}} \\ &\boldsymbol{\lambda_0 = (b_0 - 1)/a_0, \mu_0 = (b_0 - k)/a_0} \\ &\boldsymbol{a_0,\ b_0 \text{ as in } (3),\ \beta \text{ as in } (1)} \end{aligned} \tag{4}$$

The estimations based on this model, for the period 2005–2015, appear to be very close to the annual averages calculated using the NASA data [3] for the same period. According to (4), the rise in global average temperature will be estimated at

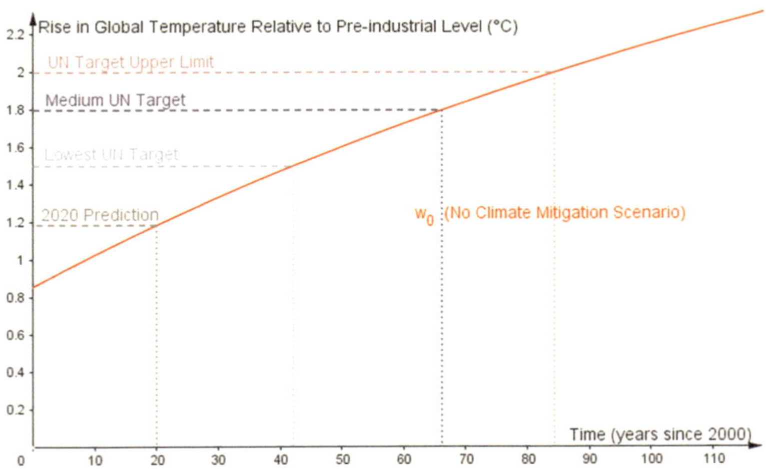

Figure 1.
Estimated global warming since 2000 and expected trend in case of no global climate mitigation.

1.19°C by 2020. Then, under the assumption of no climate mitigation, it will reach 1.5°C by 2041 and 2°C by 2084. More generally, any climate target $\mathcal{L}$ $(\mathcal{L} < \beta)$ will be hit by the year $2000 + [h]$ where h is the hitting time:

$$h \approx \frac{\mu_0 \mathcal{L} - \lambda_0 \beta}{\beta - \mathcal{L}} \qquad (\lambda_0, \mu_0 \text{ as in } (4)) \tag{5}$$

which makes the year $d = 1999 + [h]$ the deadline for the implementation of the *2015 Paris Agreement*. See **Figure 1** for graphical estimation of h.

For the rest of the paper, $\mathcal{L}$ denotes any UN climate target $(1.5 \leq \mathcal{L} < 2)$, and h is its hitting time as formulated in (5).

4. Smooth pathways for CO_2 emissions to achieve the UN goal on climate change

The consistency of future CO_2 emissions with a prescribed UN target and their rapid reduction, as urged by the UN, are crucial in the elaboration of suitable pathways for the emissions. Prior to the modelling, however, two parameters need to be determined; an estimation of their level in the beginning of the mitigation and their expected cumulative amount during the mitigation (remaining CO_2 budget).

4.1 Remaining CO_2 budget

By definition, the CO_2 budget is the total amount of cumulative anthropogenic CO_2 emitted in the atmosphere since the industrial revolution up to the time h when the UN climate target will be hit. To estimate the remaining budget at any time, future emissions need to be modelled explicitly with time in the scenario of no climate policy, which can be done by linear regression of the annual gas emissions since 2000 using CDIAC database [16]. This leads to the following no-climate-policy model E_0 (in $GtCO_2$), applicable from the year 2000 (i.e., $t = 0$):

$$E_0(t) \approx \alpha_0 t + \beta_0, \qquad \alpha_0 \approx 0.91, \beta_0 \approx 24.47 \tag{6}$$

Such a linear regression was found to be statistically highly significant $(p < 10^{-11})$ and extremely strong $(r^2 \approx 0.98)$. As a consequence of (6), the remaining CO_2 budget $R(t)$, from time t $(0 \leq t < h)$, consistent with the $\mathcal{L}$-target, is estimated as follows:

$$\boldsymbol{R(t) \approx (h - t)(\alpha_0(h + t) + 2\beta_0)/2}, \quad \boldsymbol{h} \text{ as in } (5), \boldsymbol{\alpha_0, \beta_0} \text{ as in } (6) \tag{7}$$

Indeed, with no climate mitigation, the CO_2 emissions between times t and h would reach a total amount of $R(t) = \int_t^h E_0(x)dx$, which is nothing else but the area of a trapeze with bases $E_0(t)$ and $E_0(h)$ and height $h - t$, and this gives (7).

In particular, the remaining CO_2 budgets from 2020, to meet the targets 1.5 and 1.8°C, will be estimated at 1155 and 2929 ($GtCO_2$) respectively, and these represent about 63 and 81% of the corresponding remaining budgets from 2000.

4.2 CO_2 emissions pathways consistent with the UN climate targets

One obvious way to regularly reduce the CO_2 emissions would suggest a constant rate of reduction, which will definitely put an end to them at time $t = z$

(limiting therefore the rise in global temperature to the UN target $\mathcal{L}$), as described by the following piece-wise linear model, applicable from time $t = t_0 < h$:

$$\boldsymbol{L(t) \approx \begin{cases} \dfrac{E_0}{t_0 - z}(t - z), & t_0 \leq t < z \\ 0, & t \geq z \end{cases}} \tag{8}$$

where $z = t_0 + 2R/E_0$ (with $E_0 = E_0(t_0),\ R = R(t_0)$) is determined by solving the remaining CO_2 budget equation:

$$\int_{t_0}^{z} L(t)dt = (z - t_0)E_0/2 = R. \tag{9}$$

Nevertheless, the zero-emission ensured by the linear pattern will probably cause an environmental issue, as it could not be hit before the 2070s, for UN targets as low as 1.5°C, or late 2150s for even medium targets such as 1.8°C. In addition, a constant rate of reduction (annually 31% for 1.8°C and 79% for 1.5°C) will presumably not be compatible with a struggling switch to no-/low-carbon energy.

To avoid the issue regarding the zero-emission, one could simply bring these emissions as close as possible to zero by considering a smooth non-linear pathway with an asymptotic behavior, such as the following power model:

$$\boldsymbol{P(t) \approx E_0.(t_0/t)^p, \quad t \geq t_0, \quad p = 1 + t_0E_0/R, \quad E_0 = E_0(t_0), \quad R = R(t_0)} \tag{10}$$

where the suitable power p, with $p > 1$ for integrability over $[t_0, \infty]$, that makes the model fit the given UN target, is determined by solving (for p) the associated remaining CO_2 budget equation:

$$\int_{t_0}^{\infty} P(t)dt = E_0\ t_0^p \int_{t_0}^{\infty} t^{-p}dt = t_0E_0/(p - 1) = R. \tag{11}$$

Unfortunately, the emissions could not be made as low as 0.1 ($GtCO_2$) even for the 1.5°C target and after six centuries of reduction.

However, an exponential decrease of CO_2 emissions would not only ensure their rapid reduction (as recommended in the 2015 Paris Agreement, Art. 4), but will also tolerate very low emissions, relatively earlier (compared with the power model P), that will disappear in far future, which could maintain food production, especially in the regions where transition to no-carbon energy might be extremely challenging. This leads to the following 1-phase model for CO_2 emissions consistent with a prescribed UN climate target, applicable from time $t = t_0 < h$:

$$\boldsymbol{E_1(t) \approx E_0e^{-\alpha_1(t - t_0)}, \quad t \geq t_0, \quad \alpha_1 = E_0/R, \quad E_0 = E_0(t_0), R = R(t_0)} \tag{12}$$

Indeed, to ensure an exponential decrease of the annual amount of CO_2 emissions from the initial level E_0, the model E_1 must satisfy the initial-value problem:

$$dE_1/dt = -\alpha_1E_1(\alpha_1 > 0), \quad E_1(t_0) = E_0 \tag{13}$$

which unique solution is in the form given in (10). Now, from the remaining CO_2 budget equation:

$$\int_{t_0}^{\infty} E_1(t)dt = R, \tag{14}$$

it follows that $E_0/\alpha_1 = R$, which gives α_1 as in (12).

Although the model E_1 appears to better fit the UN climate goal, in comparison with the linear and power pathways, the abrupt reduction of CO_2 emissions may threaten fundamental industries such as food production. To overcome this risk, an alternate parametrized model $E_\gamma(0 < \gamma < 1)$ would start with a succession of two smooth parabolic junctions (ascending then descending) between the linear growth and the exponential decline as follows:

$$E_\gamma(t) \approx \begin{cases} A_1(t-u)^2 + B, & t_0 \le t < u \\ A_2(t-u)^2 + B, & u \le t < v \\ E_0\ e^{-\alpha(t-v)}, & t \ge v \end{cases} \tag{15}$$

$$\begin{aligned} &\alpha = E_0/\gamma R, \quad R = R(t_0) \\ &A_1 = -\alpha_0/2\epsilon, \quad A_2 = -\alpha E_0/2\delta, \quad B = E_0 + \alpha_0\epsilon/2, \quad \alpha_0 \text{ as in } (6) \\ &u = t_0 + \epsilon, \quad v = u + \delta, \quad \epsilon = \alpha\delta E_0/\alpha_0 \\ &\delta = \left(-b + \sqrt{\Delta}\right)/2a, \quad \Delta = b^2 + 12a\alpha_0(1-\gamma)R \\ &a = \alpha E_0(\alpha E_0 + \alpha_0), \quad b = 3\alpha E_0(E_0 + 1) \end{aligned} \tag{16}$$

The free parameter $\gamma\,(0 < \gamma < 1)$ is introduced to split the remaining budget into two parts; one (γR) will go for the exponential reduction and the other $((1-\gamma)R)$ for the quadratic transition requiring two consecutive time periods; ε to slow the emissions, then δ for their initial reduction. Let $u = t_0 + \epsilon$, $v = u + \delta$ be the ending times of these periods. The coefficient α can be determined the same way as α_1 in (12) using $\int_v^\infty E_\gamma(t)dt = \gamma R$ (instead of R). On the other hand, the coefficients A_1, A_2, and B follow immediately from (2) applied with $(x_0, y_0) = (t_0, E_0)$, $m = \frac{dE_\gamma}{dt}\big|_{t=t_0} = \frac{dE_0}{dt}\big|_{t=t_0} = \alpha_0$ (for a smooth slowdown) to get A_1 and B, as given in (16), then with $(x_0, y_0) = (v, E_0)$, $m = \frac{dE_\gamma}{dt}\big|_{t=v} = -\alpha E_0$ (for a smooth transition to the exponential decline) to get A_2 as in (16) and another formulation of B $(B = E_0 + \alpha\delta E_0/2)$. Then, by equating the two expressions of B, one gets the announced formula for ϵ. As for δ, it is found to be the unique positive solution of the following quadratic equation:

$$ax^2 + bx - 3\alpha_0(1-\gamma)R = 0 \tag{17}$$

which discriminant is given by: $\Delta = b^2 + 12a\alpha_0(1-\gamma)R$, where a and b are as announced with (16). Indeed, by evaluating the integrals in the remaining CO_2 budget equation:

$$\int_{t_0}^{\infty} E_\gamma(t)dt = \int_{t_0}^{u} E_\gamma(t)dt + \int_{u}^{v} E_\gamma(t)dt + \int_{v}^{\infty} E_\gamma(t)dt = R \tag{18}$$

one gets:

$$\left((A_1/3)\epsilon^3 + B\epsilon\right) + \left((A_2/3)\delta^3 + B\delta\right) + (\gamma R) = R \tag{19}$$

Then by plugging the expressions of A_1, A_2, B, and ϵ into (19) then simplifying, one gets the following quadratic equation in δ:

$$\left(\alpha^2E_0^2/(3\alpha_0) + \alpha E_0/3\right)\delta^2 + \left(E_0 + \alpha E_0^2/\alpha_0\right)\delta - (1-\gamma)R = 0 \tag{20}$$

or equivalently,

$$\alpha E_0(\alpha E_0 + \alpha_0)\delta^2 + 3\alpha E_0(E_0 + 1)\delta - 3\alpha_0(1-\gamma)R = 0 \tag{21}$$

which gives (17) with $x = \delta$.

On the other hand, $\Delta > 0$, and more precisely $\Delta > b^2$, and hence δ, as given in (16), is the unique positive solution of Eq. (17).

Based on the model formulated in (15), **Table 1** provides an estimation of the expected reduction (in % below the 2000 level) of CO_2 emissions, due to a smoothly-implemented exponential mitigation starting by 2020, considering two different years (2050 and 2100), two climate targets (1.5 and 1.8°C), and two γ values (0.4 and 0.6). For example, whereas the emissions consistent with the 1.8°C target (for both γs) will be still above the 2000 record in 2050, the 1.5°C-pathway projects, for the same year, their reduction by 39% for $\gamma = 0.6$ and by 46% for $\gamma = 0.4$. However, the latter predicts for 2050 a similar reduction as half of the IPCC-1.5°C scenarios (70–90% below 2010 record, [5]) for $0.11 \le \gamma < 0.59$ (long to medium transition), and as the other half (95% or more below 2010 record, [17, 21]), for $\gamma < 0.11$ (long transition). More generally, as it can be seen from the rate of decline $(-\alpha E_0)$, with α as in (16), a lower target or a longer transition will require a faster reduction. See also **Figure 2** for the effect of transition length.

4.2.1 Notes

i. The parameter γ represents the fraction of the remaining budget to be used during the exponential reduction of CO_2 emissions. Consequently, the remaining fraction $(\gamma - 1)$ will go for the transition period. Therefore, the closer to 1γ is, the less CO_2 will be emitted during the transition, and the shorter the transition will be; about 14 years (resp. 4 years) with $\gamma = 0.9$, compared to about 25 years (resp. 7 years) with $\gamma = 0.8$, for the climate target 1.8°C (resp. 1.5°C). In the limit case where $\gamma = 0$ (no reduction because of no climate mitigation), the model E_γ degenerates into the linear pathway E_0 (given in (6)). However, in the other limit case where $\gamma = 1$ (no transition), the model is simply reduced to the exponential pathway E_1 (given in (12)).

ii. For all pathways E_γ, the two transition phases cannot have the same duration, i.e., there is no parameter γ for which $\epsilon = \delta$. Indeed, if this were the case, one would necessary have $\gamma = E_0^2/(\alpha_0 R)$, and this would imply that, for any climate target $\mathcal{L}$, the corresponding remaining budget R (which is an increasing function of $\mathcal{L}$) would be bounded below (by the constant E_0^2/α_0).

UNCT (°C)	R ($GtCO_2$)	Reduction by 2050 (%)[a]		Reduction by 2100 (%)[a]	
1.8	2929	None[b]	(None)[c]	35.2[b]	(46.2)[c]
1.5	1155	39.0[b]	(45.8)[c]	97.2[b]	(99.5)[c]

[a]*Reduction (in %) of CO_2 emissions below 2000 level.*
[b]*40% of remaining budget to be used for transition to exponential decline ($\gamma = 0.6$).*
[c]*60% of remaining budget to be used for transition to exponential decline ($\gamma = 0.4$).*

Table 1.
Remaining CO_2 budget (R) from 2020 and projected reduction of CO_2 emissions for 2050 and 2100 due to a global mitigation (starting by 2020) consistent with the UN climate target (UNCT).

DOI: http://dx.doi.org/10.5772/intechopen.88961

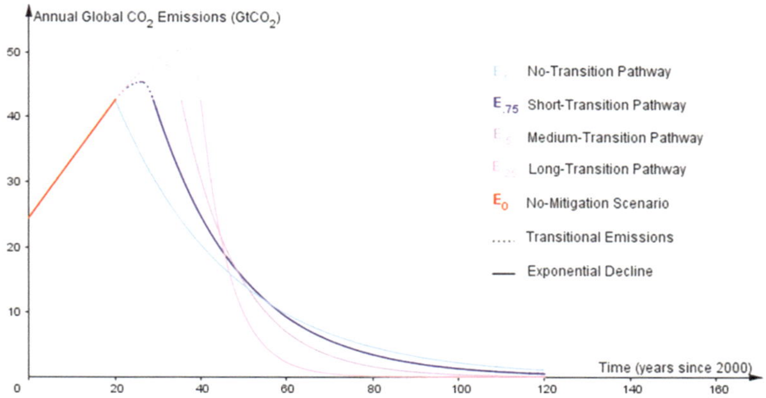

Figure 2.
CO_2 emissions pathways consistent with the UN climate target 1.5°C (mitigation starting by 2020).

Pathway E_γ	Slowdown (years)	2080	2090	2100
$\gamma = 1$	0	4.7	3.2	2.2
$.7 \leq \gamma \leq .9$	3–7	3.1–4.3	1.8–2.8	1.1 − 1.9
$.4 \leq \gamma \leq .6$	10–14	0.8–2.4	.3 − 1.3	.1 − .7
$.1 \leq \gamma \leq .3$	16–21	$3(10^{-5}) - .9$	$8(10^{-7}) - .08$	$2(10^{-8}) - .04$

Table 2.
Projected level of CO_2 emissions ($GtCO_2$) consistent with the 1.5°C target by the end of the current century (global mitigation starting by 2020).

But this would be in contradiction with the case when $\mathcal{L} \leq 1.65$, for which $R < 2001 < E_0^2/\alpha_0$. More generally, for any pathway E_γ, the times it will take the two transition phases cannot be proportional. This is due to the fact that their ratio δ/ϵ is a non-constant function of the climate target $\mathcal{L}$, as it can be seen from the formula: $\epsilon/\delta = E_0^2/(\alpha_0 \gamma R)$. This same formula shows that, for a long transition $(0 < \gamma < 0.3)$, the first phase (quadratic slowdown) will be much longer than the second (quadratic reduction). However, for a short transition $(0.7 < \gamma < 1)$, the first phase will be longer for low targets (e.g., 5 vs. 2.3 years, for 1.5°C and $\gamma = 0.8$) but shorter for higher targets (e.g., 11.5 vs. 13.4 years, for 1.8°C and $\gamma = 0.8$). The case of the 1.5°C target is illustrated in **Table 2** and **Figure 2**, where one can see that the longer the transition (decreasing γ) the longer the slowdown (increasing ϵ) and the higher the peak of emissions (coefficient B).

iii. Whereas the model E_γ for the 1.8°C target seems to be close to the (IPCC) RCP4.5, the 1.5°C version appears to be similar to the (IPCC) RCP2.6 and no- and low-overshoot over the first 30 years (see **Figure 2**), with low emissions ($<1\ GtCO_2$), e.g., by 2080 for $\gamma < 0.43$ (see **Table 2**), the same way as the no-overshoot and half of the RCP2.6 models [17, 21], or earlier, e.g., by 2050 for $\gamma < 0.08$) with the advantage of predicting the nearly-zero emission ($<0.01\ GtCO_2$), e.g., by 2090 for $\gamma < 0.22$ (see **Table 2**), or even as early as 2050 for $\gamma < 0.03$.

4.3 Ideal smooth pathways for CO_2 emissions consistent with a prescribed UN climate target

Whereas the model E_γ $(0 < \gamma < 1)$ is designed to fit a prescribed UN climate target $\mathcal{L}$°C $(1.5 \leq \mathcal{L} < 2)$, in the sense that the cumulative CO_2 emissions will not raise the global average temperature by more than $\mathcal{L}$°C above the pre-industrial level, its consistency turns out to be higher for more binding targets. Indeed, a numerical investigation (**Table 3**), based on specific criteria, shows that the lower the target the better the fit with the target. More precisely, there are (uncountable) many more E_γs, compatible with lower targets, that peak below 2.05 times (if not twice) the 2000 record (Criterion C_1) project a reduction by at least 50% in 2050 (relative to 2000 level) (Criterion C_2) and predict nearly-zero emission (≤ 0.01 $GtCO_2$) by 2100 (Criterion C_3). From an analytical point of view, what is said about criterion (C_1) is due to the decrease of the peak (coefficient B in the model E_γ) with decreasing remaining CO_2 budget R, and therefore with decreasing target $\mathcal{L}$. As for the comparison based on (C_2) and (C_3), this can be explained by the fact that the exponential decline will start more rapidly for a lower remaining budget R (due to a lower target), as it can be seen from the formula of the initial speed of the exponential reduction (at time $t = v$):

$$|\frac{dE_\gamma}{dt}| = \alpha E_0 = E_0^2/(\gamma R) \quad (22)$$

(using the formula of α given in (16)).

Consequently, the E_γs ($.24 < \gamma < .27$) appear to be the most consistent smooth pathways with the 1.5°C target, and among these, the $E_{.26}$ would be the ideal one as it satisfies the three criteria and predicts the lowest peak of emissions (by 2037), with a (constant) relative rate of exponential decline estimated at $\alpha \approx 14.2\%$. However, for more binding climate targets such as 1.4°C, which corresponding global mitigation needs to be implemented before 2034 (according to (5)), it turns out that half of the models E_γ, namely those with $0.01 < \gamma < 0.51$, meet all criteria to ideally fit this target, and the ideal one of them would be the $E_{.5}$, for the same reason as the $E_{.26}$ for the 1.5°C target, with a peak of emissions by 2028 and a (constant) relative rate of exponential decline estimated at $\alpha \approx 11.8\%$. See **Figure 3** for graphical illustration.

Nevertheless, if the 'zero' emission timing is prioritized over the peaking threshold, it is found that, among the E_γs, those with $\gamma < 0.00067$ (resp. 0.00019) project the earliest 'zero' emission; by 2043 (resp. 2035) for the 1.5°C (resp. 1.4°C) target, with a peak estimated at 2.14 times (resp. twice) the 2000 level. But the interval between the peaking and the almost-zero moments seems to be extremely short; 2 months and 10 days respectively, making the curve look like a vertical line over this interval, which sounds rather unrealistic. However, feasible ideal E_γs for the earliest 'zero' emission could be found by considering higher values of parameter γ and time z (as close as possible to the unrealistic 'zero' emission moment, as found previously) that satisfy the following conditions:

$$z - u > \max(\delta, \epsilon/3), \quad E_\gamma(z) < 0.01 \quad (23)$$

UNCT (°C)	(C_1)-Pathways[a]	(C_2)-Pathways[b]	(C_3)-Pathways[c]
1.6	$\gamma > 0.55$ (0.46)	None	None
1.5	$\gamma > 0.37$ (0.24)	$\gamma < 0.34$	$\gamma < 0.27$
1.4	$\gamma > 0.01$ (All)	All	$\gamma < 0.51$

[a]*Pathways E_γ for CO_2 emissions peaking below twice (2.05 times) 2000 level.*
[b]*Pathways E_γ for CO_2 emissions reduced by at least 50% in 2050 (rel. to 2000 level).*
[c]*Pathways E_γ for CO_2 emissions reduced to almost zero (below 0.01 $GtCO_2$) by 2100.*

Table 3.
Ideal smooth pathways for CO_2 emissions by UN climate target (UNCT) (global mitigation starting by 2020).

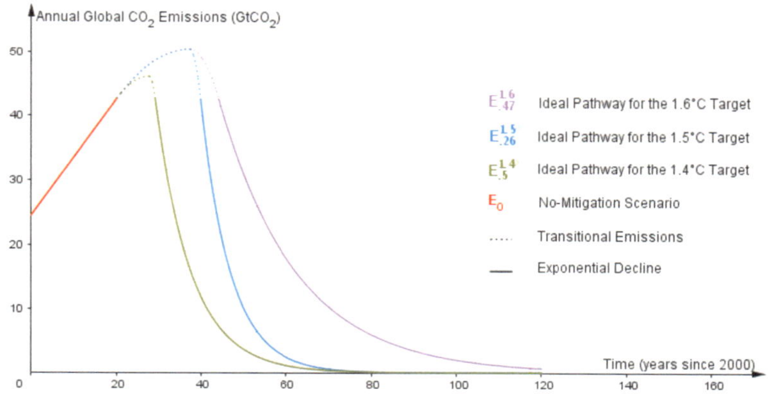

Figure 3.
Ideal smooth pathways for CO_2 emissions peaking below 2.05 times (if not twice) 2000 level and shrinking below 0.01 $GtCO_2$ by 2100 (preferably); climate targets 1.6, 1.5 and 1.4°C (mitigation starting by 2020).

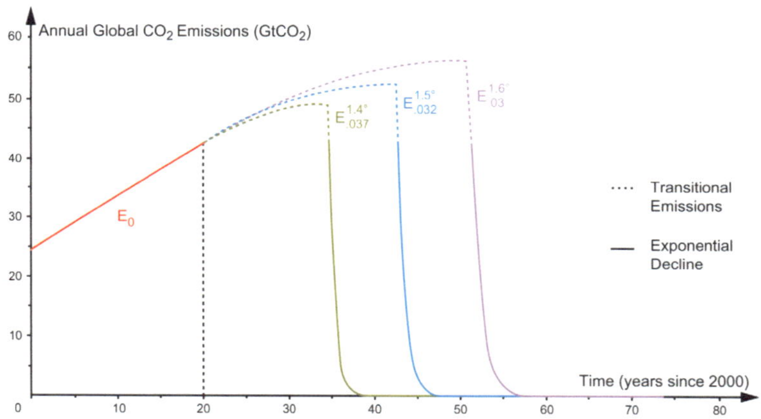

Figure 4.
Ideal smooth pathways, for the earliest feasible 'zero'-CO_2-emission; climate targets 1.6, 1.5 and 1.4°C (mitigation starting by 2020).

As a result, the earliest feasible almost-zero emission will occur by 2061, 2050, and 2040 for the respective climate targets 1.6, 1.5, and 1.4°C, with the E_γs induced by $.0179 < \gamma < .0301$, $.0163 < \gamma < .033$, and $0 < \gamma < .038$ respectively. **Figure 4** shows the most feasible of these pathways, with time periods between the peaking and the almost-zero moments estimated at 11, 8, and 6 years respectively.

An alternative ideal pathway could be made by juxtaposing the lowest restrictions of E_1 and one of the ideal E_γs consistent with the same climate target. But the resulting pattern would include two singularities (cusp shape); one in the beginning of the mitigation and another at the junction between E_1 and E_γ.

5. Conclusions

Smooth pathways for CO_2 emissions are designed taking into consideration not only their consistency with the UN climate targets but also the rapidity that has been urged by the UN for their reduction. Unlike the existing models, mostly produced by computer simulation such as the (IPCC) RCPs, a mathematical modelling, as an ideal host of interpolation and smoothing techniques, is presented throughout this chapter. First, the global warming is quantified with time to determine the moment when a prescribed UN climate target will be hit (in case of no climate mitigation), which is then used to explicitly determine the remaining CO_2 budget; crucial parameter in emissions modelling. Naturally, an exponential pattern is proposed at first for its rapid decline and long-term stabilization slightly above zero. Then, by means of quadratic interpolations, a parametrized collection of flexible pathways E_γ $(0 < \gamma < 1)$ is derived to ensure more feasibility by including a smooth transition to the exponential trend, which will help compensate a certain lack of no-carbon energy. It turns out that the no-transition (exponential) and no-mitigation (linear) models correspond to the limit values of the involved parameter γ introduced as an arbitrary fraction of the remaining CO_2 budget expected to be used during the exponential phase, which also gives an indication for the transition length.

Graphically, the E_γs are comparable to the corresponding IPCC pathways; similar to the RCP4.5, for targets between 1.5 and 2°C, and to the RCP2.6 and no- and low-overshoot, for the 1.5°C target. However, they have the advantage of predicting the nearly-zero emission (<0.01 $GtCO_2$), e.g., by 2090 for $\gamma < 0.22$, or even as early as 2050 for $\gamma < 0.03$, with no need for CO_2 removal. Such similarities could be improved by using the IPCC estimation for the remaining CO_2 budget (though determined with high uncertainties), which may lead to more representative pathways by involving further greenhouse gases.

Another virtue of the designed E_γs is their flexibility with regards to the constraints that would come with the climate target, which would provide climate policy makers with an uncountable set of ideal smooth pathways enlarging with decreasing target. For instance, whereas E_γs with $0.24 < \gamma < 0.27$ are recommended for the 1.5°C target, based on specific criteria including the peaking threshold, those with $0.01 < \gamma < 0.51$ are recommended for a more binding target; the 1.4°C one. When it comes to the projection of the earliest feasible 'zero' emission, are recommended the E_γs with $0.017 < \gamma < 0.033$ and $0 < \gamma < 0.038$, for the respective climate targets 1.5, and 1.4°C, which would result in the near extinction of CO_2 emissions by 2050 and 2040 respectively.

Acknowledgements

I am grateful to my son Seyfen for his support and growing interest in my research on global warming and its mitigation. His relevant thoughts and questions on the environment and society have been a great source of motivation to conduct this work and further projects. I especially thank him for teaching me "*Nihongo*".

Thanks also go to Yemad Mabrouk, an outstanding Math lecturer and one of my favourite colleagues, who helped resolve graphical issues.

I am also grateful to Rebekah Pribetic for her great availability as an Author Service Manager and to the whole team at IntechOpen for improving the English and editing.

Conflict of interest

The author declares no conflict of interest.

Abbreviation

CDIAC	carbon dioxide information analysis center
CO_2	carbon dioxide

$GtCO_2$	gigatons of CO_2
IPCC	Intergovernmental Panel on Climate Change
NASA	National Aeronautics and Space Administration
RCP	representative concentration pathway
UN	United Nations
UNCT	UN climate target

Author details

Nizar Jaoua
Department of Mathematics, College of Sciences and Human Studies,
Prince Mohammad Bin Fahd University, Al Khobar, KSA

*Address all correspondence to: nizar.jaoua@gmail.com

References

[1] United Nations Framework Convention on Climate Change (UNFCCC). Paris Agreement. 2015. Available from: http://unfccc.int/paris_agreement/items/9485.php [Accessed: 02 October 2016]

[2] Hawkins. Climate Spirals. Climate Lab Book. 2016. Available from: http://www.climate-lab-book.ac.uk/spirals/ [Accessed: 02 October 2016]

[3] National Aeronautics and Space Administration (NASA). Climate Change, Vital Signs of the Planet. Available from: https://climate.nasa.gov/vital-signs/temperature [Accessed: 15 September 2016]

[4] National Oceanic and Atmospheric Administration (NOAA). State of Climate, 2015 Report. 2015. Available from: http://www.noaa.gov/climate [Accessed: 08 November 2016]

[5] Intergovernmental Panel on Climate Change (IPCC). Climate Change 2014. In: The Fifth Assessment Report, AR5-SPM. Cambridge and New York: Cambridge University Press; 2014

[6] Center for Climate and Energy Solutions (C2ES). Main Greenhouse Gases. 2018. Available from: https://www.c2es.org/content/main-greenhouse-gases [Accessed: 08 September 2018]

[7] Graves C et al. New parametrizations and sensitivities for simple climate models. Journal of Geophysical Research. 1993;**198**(3):5025-5036

[8] Köse E, Kunze J. Climate modeling in the calculus classroom. The College Mathematics Journal. 2012;**44**(5): 424-427

[9] Walsh J, McGehee R. Modeling climate dynamically. The College Mathematics Journal. 2014;**44**(5): 350-363

[10] Jaoua N. Linear-fractional model for global warming. International Journal of Global Warming. 2018;**15**(2):212-226. DOI: 10.1504/IJGW.2018.092896

[11] Vuuren V et al. The representative concentration pathways: An overview. Climate Change. 2011;**109**:5-31. DOI: 10.1007/s10584-011-0148-z

[12] Knutti R. Chapter 12: Relationship between global emissions and global temperature rise. In: Climate Change 2013: The Physical Science Basis. WG1 Contribution to IPCC-AR5. 2013. 13 p. Available from: https://unfccc.int/sites/default/files/7_knutti.reto.3sed2.pdf [Accessed: 19 November 2018]

[13] Jubb I et al. Representative concentration pathways (RCPs). Australian Climate Change Science Program. 2016. 3 p. Available from: http://www.cawcr.gov.au/projects/Climatechange/wp-content/uploads/2016/11/ACCSP_RCP.pdf [Accessed: 19 November 2018]

[14] Jones CD et al. C4MIP: The coupled climate-carbon cycle model intercomparison project: Experimental protocol for CMIP6. Geoscientific Model Development. 2016;**9**(8): 2853-2880. DOI: 10.5194/gmd-9-2853-2016

[15] Rogelj J et al. Scenarios towards limiting global mean temperature increase below 1.5°C. Nature Climate Change 2018;8:325–332. DOI: 10.1038/s41558-018-0091-3

[16] Boden TA et al. Global CO_2 Emissions from Fossil-Fuel Burning, Cement Manufacture, and Gas Flaring: 1751-2013. CDIAC (Carbon Dioxide Information Analysis Center). USA:

Research Institute for Environment, Energy, and Economics; 2016. Available from: https://cdiac.ess-dive.lbl.gov/ftp/ndp030/global.1751_2013.ems [Accessed: 15 March 2019]

[17] Rogelj J et al. Chapter 2: Mitigation pathways compatible with 1.5°C in the context of sustainable development. In: Masson-Delmotte J et al., editors. Global Warming of 1.5°C. An IPCC Special Report on the Impacts of Global Warming of 1.5°C above Pre-Industrial Levels and Related Global Greenhouse Gas Emission Pathways, in the Context of Strengthening the Global Response to the Threat of Climate Change, Sustainable Development, and Efforts to Eradicate Poverty. Intergovernmental Panel on Climate Change. 2018. 82 p. Available from: https://ipcc.ch/report/sr15/ [Accessed: 07 April 2019]

[18] Manabe M, Wetherald RT. The effects of doubling the CO_2 concentration on the climate of a general circulation model. Journal of the Atmospheric Sciences. 1975;**32**(1):3-15

[19] Charney JG. Carbon dioxide and climate: A scientific assessment. Washington, D.C.: National Academy of Sciences; 1979. 22 p

[20] NASA. Climate Change, Vital Signs of the Planet. Available from: https://climate.nasa.gov/vital-signs/carbon-dioxide [Accessed: 15 September 2016]

[21] IPCC. In: Masson-Delmotte J et al., editors. Summary for Policymakers. Global Warming of 1.5°C. An IPCC Special Report on the Impacts of Global Warming of 1.5°C above Pre-Industrial Levels and Related Global Greenhouse Gas Emission Pathways, in the Context of Strengthening the Global Response to the Threat of Climate Change, Sustainable Development, and Efforts to Eradicate Poverty. Geneva: World Meteorological Organization; 2018. 32 p

Chapter 5

Reducing Green House Effect Caused by Soot via Oxidation Using Modified $LaFe_{1-x}Cu_xO_3$ Catalysts

Paritosh C. Kulkarni

Abstract

Soot has been the cause of global warming since the dawn of diesel engines. Soot oxidation in the diesel particulate filters fixated at the exhaust of the engine has been a boon to reduce the particulate matter from entering the atmosphere. Here we have attempted to synthesize the smooth powder of supported and macro-porous perovskite $LaFeO_3$ and its doped variant with 5, 10, 20, and 30% copper on B site. Subsequent catalysts, termed as LFO and LFCO- (5, 10, 20, 30) with formula $LaFe_{1-x}Cu_xO_3$, were investigated as catalysts for soot oxidation reaction. The structural and physical and chemical attributes of $LaFe_{1-x}Cu_xO_3$ and $LaFeO_3$ are characterized by XRD patterns, FESEM, BET, particle size analysis measurements. Undoped $LaFeO_3$ with desired textural structures were successfully prepared, employing the citric-acid auto combustion method, and the porous sample exhibits the best activity towards redox reactions, pointing out the enriched activity at redox sites of fabrication of porous perovskite for the responses. By correlating with the catalytic activities and the physical and chemical properties, of both doped and undoped samples, it is inferred that the best activity obtained from the porous $LaFe_{0.9}Cu_{0.10}O_3$ is attributed to its extensive surface area corresponding to least particle size, precious active lattice oxygen, high oxygen storage capacity and vigorous surface activity.

Keywords: heterogeneous catalysis, green chemistry, perovskite-type oxides, soot oxidation, sustainability, green and clean synthesis, global warming, greenhouse effect

Highlights

- Copper-doped LFO and pure LFO were analyzed as catalysts for soot oxidation.
- It is useful in oxidizing carbon particulate in comparison with pure $LaFeO_3$.
- We have correlated mechanism of soot oxidation with the function of LFO and LFCuO.
- We have proposed the optimized route to install the catalysts using the TGA chart of conversion of soot, T_{50} temperature, and catalytic performance.
- The enhanced stability and reusability indicated its potential for practical use.

1. Introduction

The literature [1] indicates a strong relationship between the rising of soot concentration in the environment and the increasing intensity of global warming over the past 20 years. Greenhouse effect and global warming are the most pressing issues the world is facing in 2019. The noticeable problem is the are average 0.17°C per year global warming since pre-industrial levels (1970–2019), also the 0.09°C warming and increasing pH of oceans 1950s, 3.2 cm sea-levels growth per decade, a large number of extreme heatwaves in last decade [2, 3]. Unless the in 2019 we act to reduce the causes of global warming, we unite and commit, and pledge that the changes to mitigate global warming are fully implemented, the adverse effects of climate change will go on.

The natural phenomenon of the greenhouse effect plays a vital role in climate change for decades. The burning of fossil fuels has contributed intensely to the natural greenhouse effect. This aggregated greenhouse effect stems from a rise in the atmospheric concentrations termed greenhouse gases. Greenhouse gases in the atmosphere lead to climate change. Soot, along with NO_x, CO, and CO_2 are considered significant greenhouse gases and pollutants.

In recent years after the industrial revolution, diesel engines are used extensively in multiple areas for their fewer CO_2 emissions, better fuel efficiency, and improved economy [4]. Soot or unburnt hydrocarbons (UHC) present in the exhaust of internal combustion diesel engines. To be in accord with the intense emission levels imposed by the law, destruction of above mentioned delirious compounds is the necessity of the era. Their conversion into a nonparticulate form before these toxic substances released into the surrounding environment is also another alternative. Soot has been counted in the chief scale of severity among such classification. For this purpose, it is proposed, the of a mixture of catalytically active components coated with perovskites will cause the parallel conversion of UHC into subsequent oxides while could be absorbed in the exhaust filters and by the environment.

While soot continues to harm human race in metropolitan cities its effect is unseen to healthy eyes, when carbon black enters your bloodstream, it can cause a wide array of severe health issues, including respiratory problems, infections, shortness of breath, bronchitis, asthma, stroke, heart attack, cancer, and premature death. Soot not only causes smog that decreases visual clarity but stays in the atmosphere longer than carbon dioxide. Carbon dioxide can be absorbed by trees, but soot cannot be 'fixed' by natural means. Soot causes global warming by absorbing sunlight and directly heat the surrounding air. **Figure 1** clearly

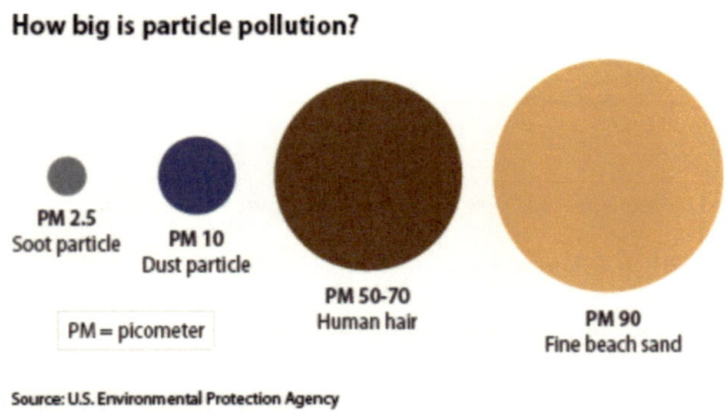

Figure 1.
Pollution caused by particulate matter.

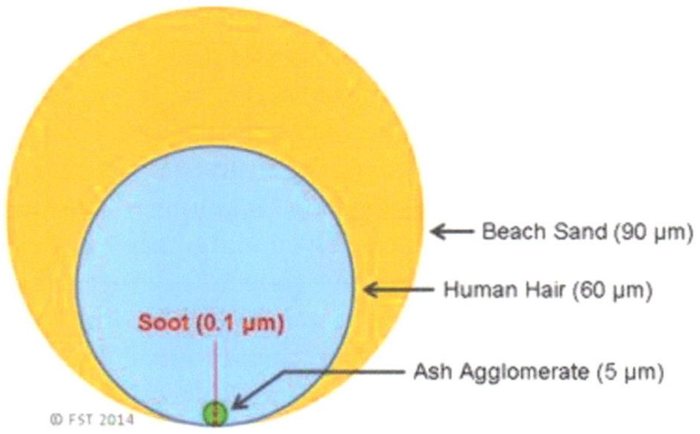

Figure 2.
Size of soot as the comparison with other particulate matter.

demonstrates how particulate matter is contributing to global warming as a significant global pollutant. **Figure 2** zones in the culmination of sizes of PM towards soot indicating inversely proportional relationship between the size of PM and its banality [5, 6].

This particulate matter (pm) in the form of soot released from the diesel combustion process has grasped plenty of attention. The reason for that is it can give rise to serious environmental and health problems [7–9]. Particulate matter is filtered with diesel particulate filters (DPF) in the current scenario [10] which captures soot and requires regeneration to maintain performance. To tackle this issue, researchers have conducted some works about regeneration, but none led to satisfaction. Methods involving in oxidation of particulate matter directly are researched upon. But they all need a high temperature above 600°C which might destroy DPF [11].

Therefore oxidation of soot is important to resolve environmental issues in many parts of the world. However, soot oxidation is not a process that can occur under natural conditions, and so further analysis is needed.

Following reactions take place for soot oxidation:

$$C_{soot} + O_2 \rightarrow CO_2 + H_2O$$

$$2CO + O_2 \rightarrow 2CO_2$$

$$C_{soot} + NO \rightarrow \tfrac{1}{2}N_2 + CO_2 + H_2O$$

As seen soot oxidation mainly involves the reaction of formation of CO_2, so does the enthalpy of reaction is enthalpy of formation. Enthalpy at 298.15 K for carbon dioxide formation is 393.5 kJ/mol and the formation of carbon monoxide is 110.53 kJ/mol. Soot oxidation is a slow process at high temperatures with relatively high activation energy.

Activation energy of above reaction at various temperature is found as follows:

$$770\text{–}1250\ \text{K}: 143.5\ \text{kJ/mol}$$

800–1000 K:161.2 kJ/mol—Accurate and depends on the partial pressure of O_2 (root pO_2)—(7)

$$1100\text{–}1400\ \text{K}: 164\ \text{kJ/mol}$$

Thus, the catalyst is required to increase the rate and decrease the time of the process. Various catalysts are proposed, and perovskites are discussed in detail. Because oxidation and reduction reactions must co-occur, perovskites are efficient owing to their high redox properties and high oxygen storage capacity (OSC) in some perovskites. In this work, we prepared a series of $LaFe_{1-x}Cu_xO_3$ perovskite-type nanopowder by sol–gel auto-combustion technique following calcination under the same experimental conditions. The metal nitrates and EDTA along with citric acid were dissolved in water, the homogeneous mixtures then added with ammonia to balance pH and kept to form a gel, and later combusted and calcined in the air at last. The as-prepared samples were characterized by XRD, FE-SEM, BET, particle size, TGA/DTA, and carbon dioxide analysis techniques to investigate the effect of the introduction amount of Cu^{2+} on the morphology, structure and redox abilities of the catalysts [12].

2. Methodology

2.1 Materials

Lanthanum nitrate hexa-hydrate ($La(NO_3)_3{\cdot}6H_2O$, ≥99.9% purity), copper nitrate ($Cu(No_3)_2$, ≥99.9% purity), and iron nitrate nonahydrate ($Fe(NO_3)_3{\cdot}9H_2O$, ≥99.9% purity) were obtained from Alfa Aesar.

2.2 Catalyst preparation

All catalyst powders were prepared using the EDTA-citrate auto combustion method [5]. Metal nitrates were employed as desired metal pre-cursors for support. A 2 g-scale preparation for $LaFe_{1-x}Cu_xO_{3-\Delta}$ is described below as an example. Lanthanum nitrate ($La(NO_3)_3{\cdot}6H_2O$) was dissolved in deionized water (100 mL), followed by mixing into an aqueous solution of copper and iron nitrates in stoichiometric ratios at room temperature. EDTA (3.8 g) dissolved in an aqueous NH_3 solution was then dropped into the mixed solution, followed by the addition of solid citric acid (3.7 g) upon stirring. Molar ratio of total metal ions (La + Fe-Cu), EDTA, and citrate is 1.0:1.0:1.5, respectively. NH_4OH was used to adjust the pH of the solution to the desired value of 11 [12–14]. The solution was then heated above 80°C slowly and became dark brown after being brown-orange at the beginning.

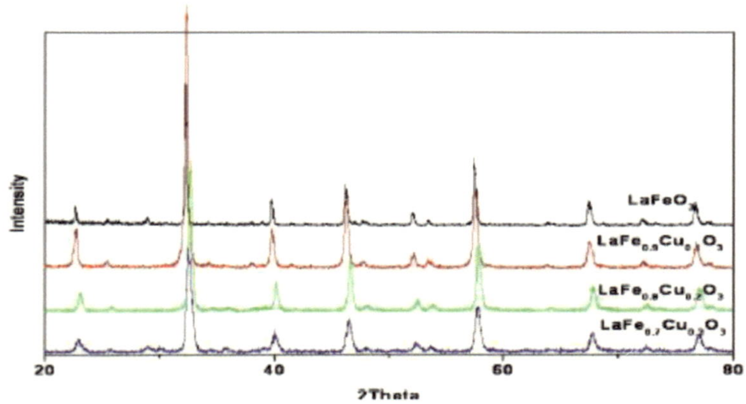

Figure 3.
XRD plot of $LaFeO_3$ doped with copper on B site.

The gel was placed in a hot air oven at 200°C for 12 h to combust and convert into flaking solid powders. Next, the powders were crushed and calcined at 800°C for 6 h in air to yield $LaFe_{1-x}Cu_xO_{3-\Delta}$ which was ground and solid power was obtained. The preparation method for undoped LFO samples is the same as above (**Figure 3**) [13].

3. Results

As shown in **Figure 4**, doping of copper of the perovskite did not affect the crystal structures of $LaFe_{1-x}Cu_xO_3$ samples (LFCO-10–LFCO-20). The peaks were in negligible deviation to the peaks of LFO. The characteristic diffraction peaks were at 22.6°, 32.2°, 38.0°, 39.6°, 46.3°, 53.3°, 57.4°, 67.4°, and 76.7° in the diffraction data of all samples can be correlated to the indices of the crystal planes of (101), (121), (112), (220), (141), (311), (240), (242), and (204), signifying that the fabricated samples were finely crystallized with three-dimensional orthorhombic structure (JCPDS No. 37-1493) [15, 16]. **Figure 4** zooms in the XRD graph of LFCO-10 and contrasts it with the pure LFO synthesized in-situ and in agreement with the JCPDS data.

Table 1 shown below depicts the a, b, c values for lattice constants of the perovskite $LaFeO_3$ and [17] $LaFe_{1-x}Cu_xO_3$. The introduction of Cu(II) with a larger ionic radius (0.730 Å) to replace Fe(III) with smaller ionic radius (0.645 Å) did not result in the expansion of LFO unit-cell [18, 19]. The smaller cell volume of $LaFe_{1-x}Cu_xO_3$ might be caused by the defects in the form of anionic vacancies, which maintained the electroneutrality in $LaFe_{1-x}Cu_xO_3$ [17, 20–26]. In **Table 1**, the crystallite sizes of Cu-doped LFO samples were smaller than that of undoped sample and decreased with increasing amount of Cu dopant.

This concurs with the literature, it shows that increasing Cu doping could cause lattice distortion and hinders the growth of large crystallites in the samples. The large degree of crystallinity with minute defects fosters the reduction in the recombination of electron–hole pairs, leading to enhanced efficiency of the soot oxidation reaction [15].

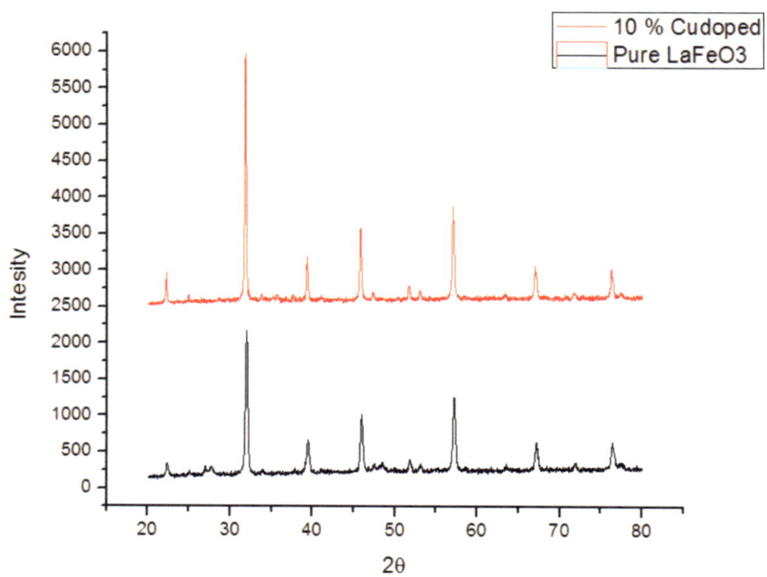

Figure 4.
Zoomed in image of XRD LFCO-10.

	a (A⁰)	b (A⁰)	c (A⁰)	Cell volume (Å^3)
$LaFeO_3$	5.576(3)	7.857(3)	5.550(2)	242.160
$LaFe_{0.9}Cu_{0.1}O_3$	5.554(4)	7.854(4)	5.554(2)	241.580
$LaFe_{0.8}Cu_{0.2}O_3$	5.567(3)	7.869(3)	5.563(2)	241.074
				240.499
$LaFe_{0.7}Cu_{0.3}O_3$	5.571(4)	7.862(4)	5.560(2)	238.101

Table 1.
Lattice parameter and cell volume calculations inferred from XRD analysis.

Sample	Structural property		
	S_{BET} (m^2/g)	Pore volume (cm^3/g)	Pore size (nm)
LFO-0Cu	17.63	0.099	45.66
LFO-5Cu	19.01	0.068	32.90
LFO-10Cu	21.37	0.11	35.52
LFO-15Cu	25.33	0.12	39.29
LFO-20Cu	24.57	0.14	37.95

Table 2.
Calculations obtained from BET analysis about structural properties.

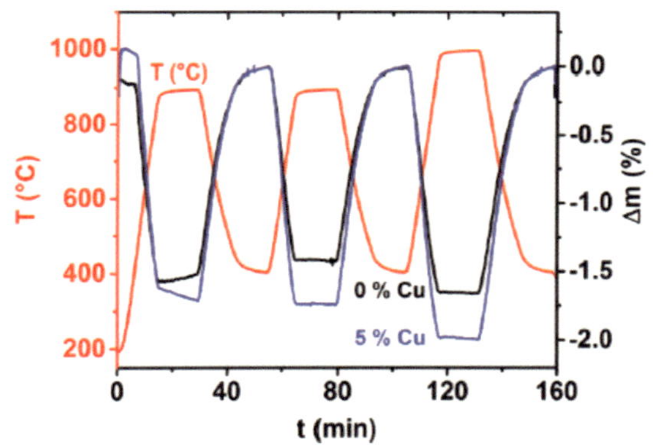

Figure 5.
TGA analysis of LFCO-5.

The information can be observed from **Table 2**, the prepared samples exhibited similar morphology, relating to the pore size consisting of nano and spherical particles. That the Cu-doping did not significantly affect the morphology and particle size of sample. The decrease in the pore size signifies that the Cu ions have been inserted into the lattice. On the other hand the XRD shoes no distortion into the lattice, which in turn explains the possible structural acceptance capability of the LFO towards favorable Cu doping.

TGA analysis results shown below proves that the catalyst is acting on the soot to reduce the carbon black particles in the controlled environment (here fixed air velocity was used) (**Figure 5**).

4. Discussion

The focus of the research is to dissect the problem of global warming by understanding scope doping of copper on the LFO structure, then to analyze the effect on the properties of the LFO and correlate them with the performance characteristics of the catalyst with the reaction of soot oxidation. Following the results, we are attempting to circumnavigate the altered properties of LFCO to the catalytical performance.

Firstly, no crystalline Cu peaks were observed and the congruent, perovskite phase formed in all cases (LFO-10Cu–LFO-30Cu). However, the amount of Cu doping in the inverse proportion with the intensity of diffraction peaks because it was broadening them.

Cell volume can be calculated for orthorhombic structure as V = abc.
For the LFCO-10-
*(since a = c), V = (5.554^2)*7.845 A°m^3 = 242.16 A°m^3 and so on*
Which gives the below inverse relation

The cell volume of $LaFe_{1-x}Cu_xO_3$ is marginally smaller than that of LFO; this value decreased with increasing Cu doping.

We can observe the trend in the BET analysis when synthesized at maximum Cu doping concentration; the particles may have fused to form large convolutions, This may also explain the more significant pore volumes and sizes for the samples of LFO-10Cu and LFO-20Cu when compared with others [8]. Increasing pore volume is the indication of more room for oxygen ions to combine with carbon, following reaction will spread light on the intricacies of the steps involved. Now as we see in the culmination of discussion, we are relating how the catalyst could have acted on the UHC (soot) and where rate-limiting step where the most unstable species carbon trioxide is formed and exactly in that step the LFO-Cu will facilitate the conversion in the diesel particulate filter (DPF) It will lower the temperature required for reaction to attain 50% conversion. From the TGA, we could infer the temperature to be around 350°C which is significantly lower than 600°C observed in reality and 500°C observed in the 0.

The controlled environment of the TGA equipment.

The chemical aspects of soot oxidation as discussed earlier indicated a need for catalyst. Here we can infer from TGA data that indeed the 50% conversion temperature of the catalyzed reaction is near to the exhaust temperature of the diesel engine, which in turn follows the possibility of soot oxidation reaction in the DPFs. Once installed the catalyst will reduce the amount of soot exiting the exhaust. The decrease in soot

$$O_{2(g)} \rightarrow O_{2(ads)} \rightarrow 2O_{(ads)}$$

$$CO_{(g)} \rightarrow CO_{(ads)}$$

$$CO_{(ads)} + 2O_{(ads)} \rightarrow CO_{3(ads)} \text{ (rate limiting step)}$$

$$CO_{3(ads)} \rightarrow CO_{2(ads)} + O_{(ads)}$$

$$CO_{2(ads)} \rightarrow CO_{2(g)}$$

Figure 6.
Possible reaction mechanism and role of catalyst—soot oxidation.

emissions from diesel engines will reduce the amount of carbon in the atmosphere, quantification of which can be extrapolated as 6 g hold per liter of diesel burnt.

As stated earlier, the formation of reactive carbon and oxidation of that species the basis of soot oxidation first order recombination reaction takes place,

$$H * + OH * + C_{(free)} - H_2O + C_{(reactive)}$$

And concerning **Figure 6**, soot oxidation mainly involves the reaction of formation of CO_2; soot oxidation is a slow process at high temperatures with relatively high activation energy (143 kJ/mol at average of 600°C) However due to increased surface activity of LFCO-20% it would be effective at the exhaust temperature itself (350°C). This would mean the catalyst is effective in converting soot to its subsequent oxides.

5. Conclusions

An attempt was made to understand the issue of global warming from all aspects, it was then established that the emission of soot contributes to global warming [1]. Soot does cause a greenhouse effect and reducing the concentration of unburnt hydrocarbons is a possible solution to combat global warming.

Soot is harmful to humans and pollutes the atmosphere, thus, soot oxidation is a necessity in today's environmental conditions, as proved soot oxidation does not take place at standard temperature and pressure and requires high temperature and stoichiometric conditions also it is slow process. Therefore, a catalyst is needed to make it feasible. To mitigate the issue of soot we undertook this research to semantically and objectively evaluate the scope doping of copper on the LFO structure, then to analyze the effect on the properties of the LFO and correlate them with the performance of the catalyst with soot oxidation.

Perovskite-type oxides were reviewed in relation to their application in soot oxidation. The main advantage of these materials is their robust crystal structure that can be used to catalyze redox reactions due to their flexible oxygen content. The possibility to accommodate simultaneously different metal cations at A- and B-sites allows tuning the catalytic properties for a specific application such as soot oxidation, which requires redox properties with high thermal stability. As such, perovskite-type oxides exhibit good oxidation activity. Cu-doped $LaFeO_3$ ($LaFe_{1-x}Cu_xO_3$) samples were prepared by a citric-acid auto combustion method and used as heterogeneous catalysts for the process of soot oxidation. The results showed that LFO-10Cu with a theoretical 10 mol% Cu doping was more effective and stable than the sample of $LaFeO_3$ (LFO) in terms of cell size and pore volumes and active surface area. The partial substitution of Cu into LFO improved oxidation rate by approximately 60%; which could be ascribed to the formation of more free oxygen during the adsorption of carbon particulates over the perovskite. The encouraging data also indicated the high stability and reusability of LFO-10Cu; therefore, it shows possible potential as a promising catalyst for soot pollutant removal in the field of diesel particulate filters [27–29]. As DPFs are optimized soot will be reduced and so global warming [1].

As of scope for further research the prepared catalyst can be manufactured at a scaled-up laboratory such facility in an industry or in the research laboratory of Pollution Control Board, Govt of India and is proposed to be implemented in public transport and custom DPFs for trials as the cost of manufacturing has to be calibrated over the years.

Acknowledgements

My Sincere thanks to Department of Chemical Engineering, Manipal Institute of Technology, Manipal Academy of Higher Education; to the faculty Dr. Harshini Dasari; and to Cintia and Luciano from IASTE.

Author details

Paritosh C. Kulkarni
Department of Chemical Engineering, Manipal Institute of Technology, MAHE, Manipal, India

*Address all correspondence to: paritosh.chandrashekhar@learner.manipal.edu

References

[1] Quaas J. Global warming: The soot factor. Nature. 2011;**471**:456-457. DOI: 10.1038/471456a

[2] Levitus S et al. World ocean heat content and thermosteric sea level change (0-2000 m), 1955-2010. Geophysical Research Letters. 2012;**39**:L10603. DOI: 10.1029/2012GL051106

[3] Climate change increases the likelihood of catastrophic avian mortality events during extreme heat waves. DOI: 10.1098/rsbl.2009.0702. Published: 30 September 2009

[4] Keav S, Matam SK, Ferri D, Weidenkaff A. Structured perovskite-based catalysts and their application as three-way catalytic converters—A review

[5] Harshini D, Lee DH, Jeonga J, et al. Applied Catalysis B: Environmental. 2014;**148-149**:415-423

[6] Simonsen SB, Dahl S, Johnson E, Helveg S. Catalytic soot oxidation studied by environmental transmission electron microscopy. In: Richter S, Schwedt A, editors. 14th European Microscopy Congress (EMC 2008); 1-5 September 2008; Aachen, Germany. Berlin, Heidelberg: Springer; 2008

[7] Labhasetwar N et al. Perovskite-type catalytic materials for environmental applications. Science and Technology of Advanced Materials. 2015;**16**:036002

[8] Fino D, Bensaid S, Piumetti M, Russo N. A review on the catalytic combustion of soot in diesel particulate filters for automotive applications: From powder catalysts to structured reactors. Applied Catalysis A: General. 2016;**509**:75-96. DOI: 10.1016/j.apcata.2015.10.016. ISSN: 0926-860X

[9] Guo H, Anderson P, Sunderland P. Optimized rate expressions for soot oxidation by OH and O_2. Fuel. 2016;**172**. DOI: 10.1016/j.fuel.2016.01.030

[10] Singh M, Gaddam CK, Wal RLV, Xu C, Yezerets A, Kamasamudram K. Reconciliation of Carbon Oxidation Rates and Activation Energies Based on Changing Nanostructure. John and Willie Leone/Department of Energy & Mineral Engineering (EME)/Institutes of Energy and the Environment (IEE)/ Materials Research Institute (MRI)

[11] Jung H, Kittelson D, Zachariah M. Kinetics and visualization of soot oxidation using transmission electron microscopy. Combustion and Flame. 2004;**136**:445-456. DOI: 10.1016/j.combustflame.2003.10.013

[12] Ovejero G, Sotelo JL, Romero MD, Rodríguez A, Ocaña MA, Rodríguez G, et al. Multiwalled carbon nanotubes for liquid-phase oxidation. Functionalization, characterization, and catalytic activity. Industrial & Engineering Chemistry Research. 2006;**45**(7):2206-2212. DOI: 10.1021/ie051079p

[13] Thirumalairajan S, Girija K, Hebalkar NY, Mangalaraj D, Viswanathan C, Ponpandian N. Shape evolution of perovskite $LaFeO_3$ nanostructures: A systematic investigation of growth mechanism, properties and morphology dependent photocatalytic activities

[14] Industrial & Engineering Chemistry Research. 2014;**53**(24):10090-10096. DOI: 10.1021/ie501277x. Publication Date: May 23, 2014

[15] Dacquin JP, Lancelot C, Dujardin C, Cordier-Robert C, Granger P. Support-induced effects of $LaFeO_3$ perovskite on the catalytic performances of supported

Pt catalysts in DeNOx applications. Journal of Physical Chemistry C. 2011;**115**(5):1911-1921. DOI: 10.1021/jp1069779. Publication Date: 06 January 2011

[16] Di Sarli V, Landi G, Lisi L, Saliva A, Di Benedetto A. Catalytic diesel particulate filters with highly dispersed ceria: Effect of the soot-catalyst contact on the regeneration performance. Applied Catalysis B: Environmental. 2016;**197**:116-124. DOI: 10.1016/j.apcatb.2016.01.073. ISSN: 0926-3373

[17] Kulkarni PC, Neelapala SD, Uppara HP, Dasari H, Van Esch N. $SmFe_{1-y}Co_yO_3$ perovskite-type oxides for soot oxidation. International Journal of Recent Technology and Engineering (IJRTE). 2019;**8**(1S4). ISSN: 2277-3878

[18] Trovarelli A, Hutchings GJ, editors. Catalysis by Ceria and Related Materials (Università di Udine, Italy). Catalytic Science Series. Vol. 2. London: Imperial College Press; 2002. vviii + 508p. ISBN: 1-86094-299-7

[19] Swartz SL. Journal of the American Chemical Society. 2002;**124**(43):12923-12924. DOI: 10.1021/ja025256e. Publication Date: September 24, 2002

[20] Xiao P, Zhong L, Zhu J, Hong J, Li J, Li H, et al. CO and soot oxidation over macroporous perovskite $LaFeO_3$. Catalysis Today. 2015;**258**(Part 2):660-667. DOI: 10.1016/j.cattod.2015.01.007. ISSN: 0920-5861

[21] Yaghi OM, O'Keeffe M, Ockwig NW, Chae HK, Eddaoudi M, Kim J. Reticular synthesis and the design of new materials. Nature. 2003;**423**:705-714

[22] Wei Y, Liu J, Zhao Z, Chen Y, Xu C, Duan A, et al. Highly active catalysts of gold nanoparticles supported on three-dimensionally ordered macroporous $LaFeO_3$ for soot oxidation. Angewandte Chemie, International Edition. 2011;**50**:2326-2329. DOI: 10.1002/anie.201006014

[23] Lukashuk L, Yigit N, Li H, Bernardi J, Föttinger K, Rupprechter G. Operando XAS and NAP-XPS investigation of CO oxidation on meso- and nanoscale CoO catalysts. Catalysis Today. 2019;**336**:139-147. DOI: 10.1016/j.cattod.2018.12.052. ISSN: 0920-5861

[24] Ma BQ, Sun HL, Gao S. The design and synthesis of a non-centric diamond-like network based NH_4^+ ion. Chemical Communications. 2003:2164-2165

[25] Mangal S et al. Materials Today: Proceedings. 2018;**5**:16378-16389

[26] Ramirez JH, Maldonado-Hódar FJ, Pérez-Cadenas AF, Moreno-Castilla C, Costa CA, Madeira LM. Azo-dye orange II degradation by heterogeneous Fenton-like reaction using carbon-Fe catalysts. Applied Catalysis B: Environmental. 2007;**75**(3-4):312-323. DOI: 10.1016/j.apcatb.2007.05.003. ISSN: 0926-3373

[27] Mishra A, Prasad R. Preparation and application of perovskite catalysts for diesel soot emissions control: An overview. Catalysis Reviews. 2014;**56**(1):57-81. DOI: 10.1080/01614940.2014.866438

[28] Li L, Wang X, Zhang Y. Materials Research Bulletin. 2014;**50**:18-22

[29] Sannino D, Vaiano V, Ciambelli P, Isupova LA. Catalysis Today. 2011;**161**:255-259

Chapter 6

Ca-Cu Chemical Looping Process for Hydrogen and/or Power Production

Isabel Martínez, Jose R. Fernández and Gemma Grasa

Abstract

It has been widely reckoned the potential of developing novel CO_2 capture technologies aiming at low-energy penalties and reduced cost as a solution for fighting against climate change. The Ca-Cu chemical looping process emerged as a promising technology for producing hydrogen and/or power with inherently low CO_2 emissions. The core of this concept is the calcination of the $CaCO_3$ by coupling in the same solid bed the exothermic reduction of a CuO-based material, improving the efficiency of the CO_2 sorbent regeneration step. Significant progress has been made since its first description in 2009, fulfilling the validation of the key stage under relevant conditions for the process in 2016. This chapter compiles the main advances in the Ca-Cu process regarding material development, reactor and process design and lab-scale testing, as well as in process simulation at large scale.

Keywords: CO_2 capture, sorption-enhanced reforming, calcium looping, chemical looping

1. Introduction

It is globally accepted that there is an unequivocal relation between the increment of anthropogenic greenhouse gas (GHG) emissions to the atmosphere and the rise in the global temperatures [1]. CO_2 is considered the principal GHG due to the magnitude of its emissions in the global scenario (i.e. about 78% of total GHG emissions in the 2000–2010 period corresponded to CO_2), having reached a value of 36.2 $Gton_{CO_2}$ in 2015 [2]. Fossil-fuel combustion is the responsible of about 90% of CO_2 emitted, being the heat and power sectors the major contributors to this share (i.e. about 35%). Among the industrial sectors with the largest CO_2 emissions, iron and steel manufacturing, cement production and other chemical industries (i.e. ammonia or lime production) are the most important. Drastic CO_2 emission reductions are needed to contribute in the stabilization of the global temperature rise to about 1.5°C above the pre-industrial levels, as recently agreed in the 22nd Conference of the Parties in 2016. In this context, CO_2 capture and storage (CCS) has raised as the only option for drastically reducing the CO_2 emissions in large stationary sources beyond the limits needed for fulfilling such ambitious target [1].

Hydrogen represents a proper alternative to fossil fuels due to its flexibility, fuel density and low carbon footprint. Currently, around 90% of the hydrogen produced worldwide (i.e. about 65 million tons per year) is used as raw material for ammonia

IntechOpen

and methanol production [3]. However, fossil fuels represent the principal feedstock for hydrogen production worldwide, with around 96% of the global hydrogen produced from natural gas, fuel oil and coal. Hydrogen production with low carbon footprint has a great potential in fulfilling the stringent CO_2 emission cuts needed in the energy sector [4]. Therefore, the development of large-scale hydrogen production technologies including CO_2 capture that enable a reduced cost as well as an improved efficiency would greatly contribute to the climate change mitigation route [5].

In this context, the steam methane reforming (SMR) coupled with in situ CO_2 separation is gaining importance as a method for obtaining high-purity hydrogen in one single step [5]. This sorption-enhanced reforming (SER) proposes carrying out the reforming of methane in the presence of a CO_2 sorbent that reacts with the CO_2 as soon as it is formed, pushing the reaction equilibrium towards hydrogen production [6, 7]. Due to its good performance and favourable kinetics, CaO-based materials have been typically proposed as CO_2 sorbents in the SER process [7]. According to SER equilibrium (see Eqs. (1)–(3)), using CaO as a CO_2 sorbent allows reaching H_2 contents as high as 96 vol.% (dry basis) in a single step for temperatures about 650–700°C [6]. No water-gas shift (WGS) reactors are needed downstream the SER process since the CO content in the syngas produced is very low thanks to the presence of the CO_2 separation process. Moreover, since all the reactions occur in a single reactor, the energy released by the exothermic CaO carbonation reaction and the WGS reaction compensates the energy required for the reforming of CH_4, resulting in an almost neutral system that does not need from an external energy source as in conventional SMR.

$$CH_4 + H_2O \leftrightarrow CO + 3H_2 \quad \Delta H_{298\,K} = 206.2 \text{ kJ/mol} \tag{1}$$

$$CO + H_2O \leftrightarrow CO_2 + H_2 \quad \Delta H_{298\,K} = -41 \text{ kJ/mol} \tag{2}$$

$$CaO + CO_2 \leftrightarrow CaCO_3 \quad \Delta H_{298\,K} = -178.8 \text{ kJ/mol} \tag{3}$$

One of the main issues of the SER process is the $CaCO_3$ regeneration step, which is a high endothermic reaction that needs to be performed continuously to allow for a cyclic operation. Several alternatives have been proposed in the literature for supplying the large amount of energy needed in the $CaCO_3$ regeneration step. Commonly, the direct combustion of a fuel in the same reactor in the presence of an O_2-rich atmosphere has been proposed [8, 9], which will allow producing a CO_2 stream that is not diluted with N_2 and so easy to be purified and compressed for its final storage. Other options have been proposed as an alternative to the high energy-consuming air separation unit (ASU) needed for supplying the pure O_2 required in this direct combustion option. For instance, the introduction of a high-temperature solid stream coming from a combustor in the calciner [10] or the use of an integrated high-temperature heat exchanger in the regenerator for transferring the heat indirectly from a high-temperature fluid [11, 12] have been proposed, but both options have not reached a sufficient development stage due to their limitations. As an alternative method for solving the problem of $CaCO_3$ regeneration in the SER process, the Ca-Cu looping process emerged [13]. This process proposes carrying out the calcination of the $CaCO_3$ by coupling in the same reactor the exothermic reduction of CuO with a gaseous fuel (i.e. containing CH_4, H_2 and CO). In this way, the coupling of the endothermic and exothermic reactions in a single step allows to supply directly the heat needed for $CaCO_3$ calcination without the need of costly heat exchange surfaces or energy-demanding units like the ASU, resulting in this way in a high-efficiency process.

2. Ca-Cu looping process: the concept

The Ca-Cu looping process was originally proposed in 2009 by Abanades and Murillo [13], and its basic scheme is based on the three main reaction stages shown in **Figure 1** [14]. The reactor configuration that fully exploits the advantages of the proposed concept is a series of fixed-bed reactors that operate in parallel at different pressure and temperature. Each fixed reactor passes through each stage of the Ca-Cu process in a sequential manner when changing the feed gas. Three functional materials are needed for operating this process: (i) a Cu-based material, (ii) a Ca-based CO_2 sorbent and (iii) a reforming catalyst (typically Ni-based).

The Ca-Cu process can be applied as a post-combustion CO_2 capture process in power plants, but the application having received more attention has been the developing of processes for the production of high-purity hydrogen and/or power [15]. At the beginning of the process, the Cu-based material and the reforming catalyst should be present in the bed in their reduced form, whereas the CO_2 sorbent should be fully calcined. The first stage of the process (referred to as 'A' in **Figure 1**) consists of a SER process, and it starts when natural gas and steam are fed to the reactor. The SMR, WGS and CaO carbonation reactions (Eqs. (1)–(3), respectively) occur during this stage. Pressure proposed to operate this SER stage ranges from about 10 to 25 bar depending on the main output of the process (i.e. hydrogen or power production). A H_2-rich gas is obtained at the outlet of this stage at high temperature, which should be cooled down to be used as fuel in a power production process or to be exported and used as feedstock for a downstream chemical process. SER stage finishes when all the CaO present in the solid bed is fully carbonated and there is no extra CaO to react with CO_2. The Cu-based material remains in its reduced form, unreacted, through this SER stage.

The second stage of the Ca-Cu process consists of the oxidation of the Cu present in the solid bed to produce the amount of CuO needed for the calcination of the $CaCO_3$. The oxidation stage (indicated as 'B' in **Figure 1**) starts when diluted air is fed to the fixed-bed reactor. This stage should be operated at controlled conditions of temperature and pressure to avoid temperature peaks within the reactor that lead to the prompt decomposition of the $CaCO_3$, as well as to avoid operational problems related to the Cu-based material (i.e. agglomeration and/or loss in reactivity) [16]. The operation strategy proposed for limiting such maximum allowable temperature has been reducing the temperature and the O_2 content in the diluted air stream fed to this stage. Recirculating a large fraction of the O_2-depleted gas at this stage outlet dilutes the O_2 content in the feed gas and increases the flow rate of the gas

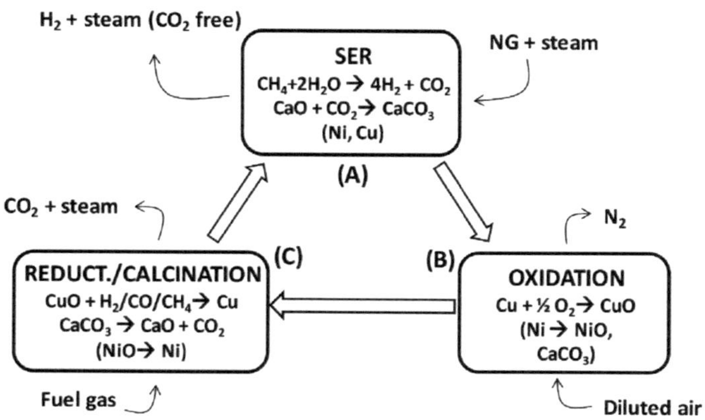

Figure 1.
Conceptual scheme of the Ca-Cu looping process.

fed to this reactor. This oxidation stage is operated at high pressure for reducing the driving force towards $CaCO_3$ decomposition. O_2 contents of around 3 vol.% and inlet gas temperatures of 150–300°C for this oxidation stage have been proposed as suitable for limiting the maximum temperature reached within the solid bed at 830–850°C [17]. The operation at high pressure allows using the non-recirculated O_2-depleted gas at the reactor outlet for producing electricity in a gas turbine. The Cu oxidation stage finishes when all the Cu present in the solid bed is oxidized into CuO. At this moment, all the solid bed has been left at the temperature of the inlet gas (i.e. around 300°C), which is too low for the subsequent calcination/reduction stage to begin. The recirculated O_2-depleted gas exiting the oxidation stage needs to be cooled down to about 300°C to be fed to the reactor, and it is passed through the fully oxidized bed to be cooled down while transferring its sensible heat to the solids, leaving them at a temperature of around 760–800°C suitable for the next reaction stage.

The last stage of the Ca-Cu process consists of the calcination of the $CaCO_3$ formed during the SER process by means of the energy released by the exothermic reduction of the CuO. This stage operates at atmospheric pressure to limit the maximum temperature needed to around 850–870°C. Typically, a mixture of H_2, CH_4 and CO is proposed as feed gas in this stage, coming from either a hydrogen purification section or a separate reforming process. Proper CaO and Cu amounts are needed in the reactor to ensure that the energy released by CuO reduction is enough for fulfilling the energy requirement from $CaCO_3$ calcination as well as to reach the desired calcination temperature. The Cu/Ca ratio needed depends on the composition of the fuel gas used in this stage considering the reaction enthalpies of the reduction reactions with H_2, CH_4 or CO (see Eqs. (4)–(6)). The maximum amount of Cu is needed when using CH_4 as reducing gas since it leads to the lowest exothermic reduction reaction per mole of CuO (i.e. a Cu/Ca molar ratio of 3.1 considering the reaction enthalpies at 850–870°C). On the contrary, the largest reduction enthalpy of the CuO with CO leads to the lowest Cu/Ca molar ratio needed of 1.3 [17].

$$4CuO + CH_4 \rightarrow 4Cu + CO_2 + 2H_2O \quad \Delta H_{298\ K} = -158.3\ kJ/mol \tag{4}$$

$$CuO + H_2 \rightarrow Cu + H_2O \quad \Delta H_{298\ K} = -85.8\ kJ/mol \tag{5}$$

$$CuO + CO \rightarrow Cu + CO_2 \quad \Delta H_{298\ K} = -126.9\ kJ/mol \tag{6}$$

3. Development of materials suitable for the Ca-Cu process

Three functional materials are needed for running the Ca-Cu process, namely, the CaO-based CO_2 sorbent, the Cu-based material and the reforming catalyst. Their proportion in bed will be determined by, on the one hand, the energy balance in the calcination/reduction stage in the case of Cu-based material/sorbent and, on the other hand, the CH_4 space velocity that a system is able to convert for the sorbent/catalyst ratio. In any case, it is important to maximize the active phase in every material, as the presence of inert in the reactor would negatively affect the efficiency of the process. In this section, a revision on the recent developments of CaO and Cu-based materials suitable for this process is included. As for the catalyst, conventional Ni-based reforming catalysts have been typically proposed for this process [15, 17], which have been successfully tested under suitable conditions for the Ca-Cu process [18, 19]. It is important to assess the effect that the redox cycles have on catalyst activity and to determine the operational window in terms of CH_4 space velocity that a system sorbent/catalyst is able to convert.

3.1 Development of Ca-based CO_2 sorbents

In the recent years, intense work has been carried out in the field of synthetic CaO-based sorbents with the objective of overcoming the decay in CO_2 capture capacity that presents CaO-based sorbents derived from natural limestones and dolomites (see e.g. recent reviews [20–22]). Among the different strategies followed to produce materials resistant to sintering, the incorporation of the active phase (i.e. CaO) into an inert matrix is the most extended and validated synthesis method [23, 24]. The performance of the materials (referred to as CO_2 carrying capacity in g_{CO_2} absorbed per unit of CaO or sorbent weight) is commonly related to their pore structure/surface area and its evolution with the reaction cycles. The decay in sorption capacity is mainly a result of a sintering phenomenon that consists of the agglomeration of small CaO grains and the evolution of the pore structure towards higher pore diameters. It is important to highlight that it is not possible to directly compare the behaviour of materials tested under diverse reaction conditions as these will affect the materials performance in the long term (high number of reaction cycles) [25–27]. Anyway, there are valid trends that can be extracted from the results in the literature, as for example, that the maximum CO_2 carrying capacity of a sorbent is directly proportional to its CaO load [20] and that a minimum amount of inert matrix is required to maintain the pore structure and so reduce sintering.

A wide variety of synthesis methods (i.e. wet mixing, spray pyrolysis, sol-gel, co-precipitation, etc.) and inert supports (i.e. Al_2O_3, MgO, ZrO_2, SiO_2, $Ca_{12}Al_{14}O_{33}$, etc.) have been studied in the literature for preparing synthetic CaO-based sorbents (see detailed reviews [20, 22] for an extended list of synthetic CaO-based sorbents). A recent paper by López et al. [24] evaluated the effect that sorbent inert support has on CO_2 carrying capacity and reactivity towards carbonation reaction. For this purpose, materials with different CaO amounts were prepared using two different inert supports (i.e. MgO and $Ca_{12}Al_{14}O_{33}$). CaO/MgO materials were prepared through co-precipitation (with CaO contents between 100 and 40%wt.), whereas materials with $Ca_{12}Al_{14}O_{33}$ as inert support were prepared via mechanical mixing and later calcination. The results indicated that a minimum amount of inert species was required to stabilize and to improve the CO_2 carrying capacity of the materials beyond the capacity of pure CaO. In **Figure 2,** the CO_2 carrying capacity of the different synthetic CaO-based materials prepared in [24] is depicted. A minimum amount of 10%wt. MgO improves the CO_2 carrying capacity of the material with respect to the performance of the co-precipitated CaO. Moreover, reducing the amount of CaO in the material diminishes the decay in the CO_2 sorption capacity along the initial cycles that is typical of naturally derived CaO-based materials. Taking into account that operation of the Ca-Cu process is thought to be carried out in fixed-bed reactors, the different functional materials should be in particle or pellet form. López et al. [24] prepared particles through an agglomeration process from the synthetized powder and demonstrated that the agglomeration process affected the textural properties of the materials, reducing the BET surface area and porosity with respect to the properties of the powder. Synthetic dolomites with a CaO/MgO molar ratio of 2:1 and a particle size cut of 0.6–1 mm were obtained, which showed a CO_2 carrying capacity of about 0.28 g_{CO_2}/g calcined material after 100 reaction cycles performing calcination under realistic conditions for the Ca-Cu process (i.e. at 900°C and 70%vol. CO_2).

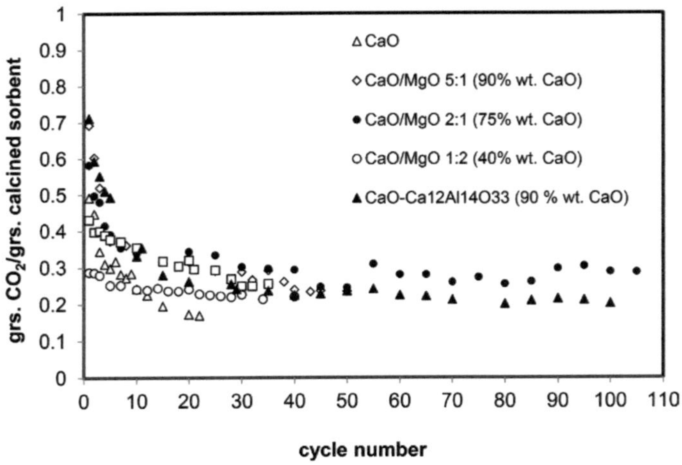

Figure 2.
Evolution of CO_2 carrying capacity with the number of cycles for different CaO-based materials (adapted from the information published in [24]). White symbols correspond to materials tested in powder form (<75 μm) and black symbols to materials in particle size cut of 0.6–1 mm.

Promising results have been reported in the literature for materials with $Ca_{12}Al_{14}O_{33}$ as inert support prepared through diverse synthesis routes under relevant calcination conditions for the Ca-Cu process (i.e. temperatures above 900°C in presence of CO_2 and steam). Pacciani et al. [28] reported a CO_2 carrying capacity of 0.17 g_{CO_2}/g calcined sorbent after 110 reaction cycles for a 85%wt. CaO, 15% wt. $Ca_{12}Al_{14}O_{33}$ sorbent prepared by co-precipitation. In another study, Koirala et al. [29] prepared different Ca-based sorbents with different Al/Ca molar ratios via single-nozzle flame spray pyrolysis. A CO_2 carrying capacity of 0.25 g_{CO_2}/g calcined sorbent was demonstrated after 100 calcination/carbonation cycles for a material with an Al/Ca molar ratio of 3:10 under severe calcination. Radfarnia and Sayari [30] used a citrate-assisted sol-gel technique followed by a two-step calcination method to produce an Al-stabilized sorbent that presented a CO_2 carrying capacity of 0.33 g_{CO_2}/g sorbent after more than 30 reaction cycles calcined at 930°C and 100% CO_2. An effort has been done by Kazi et al. [31] to produce efficient and stable CaO-$Ca_{12}Al_{14}O_{33}$ sorbents via a cost-effective and easy scalable hydrothermal synthesis route, starting from low-cost hydroxide precursors. Through this method a highly stable sorbent presenting 0.21 g_{CO_2}/g calcined sorbent was synthesized, whose production has been recently scaled up within the framework of the FP7 ASCENT project [32].

3.2 Development of Cu-based materials

There is an important number of works focused on the development of Cu-based materials due to their application in chemical looping processes as oxygen carriers [16]. Different synthesis routes have been reported in the literature for these materials, as freeze granulation, impregnation, extrusion, spray drying, co-precipitation or mechanical mixing, using different inert supports (i.e. Al_2O_3 as the most common, but also $MgAl_2O_4$, ZrO_2, CeO_2, TiO_2 and SiO_2 have been proposed), as widely reviewed by Adánez et al. [16]. Cu-based materials with high Cu loads (i.e. about 60%wt. Cu) highly resistant to agglomeration and deactivation are those suitable for the Ca-Cu process. The recent interest of combusting solid fuels through CLOU process speeded up the development of materials with higher oxygen transport capacity (OTC) and therefore higher Cu contents [33]. However, despite some works reporting stable OTC along a reduced number of cycles operated in fluidized bed under CLOU conditions for materials containing 80%wt. CuO on to $MgAl_2O_4$ [34], there are not many works published so far about highly loaded Cu materials specifically prepared for operation in fixed-bed reactors in pellet or large-particle form. A recent paper from Díez-Martín et al. [35] evaluated the maximum CuO load onto different inert supports (Al_2O_3, $MgAl_2O_4$, ZrO_2) that allowed chemically and mechanically stable materials along representative conditions for the Ca-Cu process. According to the results from this work, it was possible to produce chemically and mechanically stable pellet materials with Cu contents up to 65% wt. onto Al_2O_3 and $MgAl_2O_4$ from co-precipitated powders.

3.3 Development of combined CaO-CuO materials

With the objective of improving heat and mass transfer phenomena within the reduction/calcination stage of the process, there is an increasing number of works evaluating the synthesis of combined functional Ca-Cu materials [36–39]. Mechanical mixing of CaO and CuO powders was the selected synthesis route followed by Manovic and Anthony [40] for synthetizing for the first time this combined material. These authors prepared pellets by mixing CaO from calcined natural limestone with commercial CuO particles and Ca-aluminate cement as binder in a proportion that resulted in 45:45:10 mass fraction. Material performance was evaluated in a TGA apparatus along successive reduction/calcination and oxidation cycles. The Cu phase was totally converted during reduction (at 800°C in a CH_4 atmosphere) and oxidation in air, indicating that this could be a suitable material for the Ca-Cu process. Trying to explore the possibilities of the synthesis route, the same authors prepared core-in-shell materials with different CaO, CuO and Ca-aluminate cement proportions [41], maintaining the CuO in the inner core of the pellet. The OTC of the pellets indicated that a 25%wt. CaO in the core is sufficient to support the CuO and prevent the decay of its activity as an oxygen carrier. In other works, Quin et al. [39, 42] assessed the performance of materials composed by CaO and CuO supported on to MgO, Al_2O_3 or cement, prepared following diverse synthesis routes (wet mixing, sol-gel and mechanical mixing). The materials were tested in TGA and showed good reactivity along reduction and oxidation cycles using CH_4 and air, respectively. The presence of inert support allowed the combined material to maintain its CO_2 carrying capacity along cycles. This was especially clear for materials containing MgO on its structure, as this species greatly reduced the resistance to CO_2 diffusion during the carbonation stage. In contrast, the presence of Al_2O_3 produced $Ca_{12}Al_{14}O_{33}$ after reaction with CaO reducing in this way the amount of active phase for the carbonation reaction. As it happened for the sorbent and oxygen carriers, co-precipitation has been also a synthesis route explored to produce combined materials. Kierzkowska and Müller [38] prepared through this route combined materials with diverse CaO and CuO contents (CaO:CuO molar ratios of 1:1, 1.3:1 and 3.3:1) to be tested in a TGA along multiple carbonation/reduction/calcination/oxidation cycles. These cycles were performed isothermally at 750°C, carrying out carbonation in 40%vol. CO_2 in air, reduction in 10%vol. CH_4 in N_2 and oxidation 4.2%vol. O_2 in N_2. According to this study, the best result was obtained for the material with a molar ratio 1:1 that maintained a CO_2 carrying capacity of 0.18 g_{CO_2}/g material after 15 reaction cycles. In every case, the Cu phase reacted over 98%. Also these authors explored the effect that inert species might have on the chemical stability of the combined material, in this way they prepared via sol-gel materials supported on to Al_2O_3, MgO and $MgAl_2O_4$ with CuO and CaO molar ratio of 1.3:1 and 3.3:1 [43]. As found by other authors, the presence of Mg in the support stabilized the CO_2 uptake and minimized carbon deposition. CuO-$MgAl_2O_4$ with a proportion of 1.3:1 was the material with the highest CO_2 uptake of 0.13 g_{CO_2}/g material after 15 cycles of repeated carbonation/

calcination-redox reactions. In line with the efforts made to produce effective and economic sorbent materials, Kazi et al. [44] developed combined Ca-Cu materials via the hydrothermal synthesis route. A CO_2 carrying capacity of 0.15 g_{CO_2}/g material and an oxygen transport capacity of 0.07 g_{O2}/g material after 50 reaction cycles in a TGA were reported for a material composed of 53%wt. CuO and 22% wt. CaO, being the rest $Ca_{12}Al_{14}O_{33}$. Conditions used for TGA tests were carbonation using a gas mixture of 15%vol. CO_2, 25%vol. steam in N_2 at 650°C, oxidation at 870°C with 25%vol. air in CO_2 and reduction at 870°C in a 40%vol. CO_2, 25% vol. steam in N_2.

4. Ca-Cu process lab-scale testing

The feasibility of the reaction steps of the Ca-Cu looping process has been experimentally confirmed in packed-bed reactors at laboratory scale during the recent EU-FP7 Project ASCENT [32]. Grasa et al. [18] focused the investigation on the SER stage using a commercial Ni-based catalyst and a $CaO-Ca_{12}Al_{14}O_{33}$ sorbent. After 200 reduction/oxidation cycles, the sorbent/catalyst system produced a gas with more than 90 vol.% H_2 on a dry basis (i.e. close to the maximum equilibrium value), operating with space velocities up to 2.5 kg CH_4 h^{-1} kg cat^{-1} (i.e. a gas velocity of 0.53 m/s inside the bed). The maximum space velocity at which the CH_4 is totally converted during the SER operation is determined by the CaO carbonation reaction. Sorbent carbonation reaction rates up to 4.42×10^{-2} kmol h^{-1} kg $sorbent^{-1}$ were calculated in the experiments.

The feasibility of the Cu oxidation stage was experimentally demonstrated by Alarcón et al. [45] under relevant conditions for the Ca-Cu looping process. Oxygen in the feed was diluted to 3%vol. with N_2 simulating the recirculation of a large fraction of the product gas from the oxidation stage outlet. The maximum temperature in the bed was kept below 800°C during the oxidation, which should prevent the agglomeration or sintering of the Cu-based material and highly reduce the loss of CO_2 by the partial calcination of the sorbent. Even at low starting temperatures in the reactor (of about 400°C), the oxidation of Cu occurred very fast taking place in sharp reaction fronts throughout the reactor. During the pre-breakthrough period, complete conversion of O_2 was observed despite of the very low O_2 content in the feed.

Fernández et al. [46] demonstrated at TRL4 the viability of the calcination of $CaCO_3$ by the in situ reduction of CuO with H_2 giving rise to a product gas composed of virtually pure CO_2 (after the condensation of H_2O). Tests were carried out in a fixed-bed reactor (1 m long and inner diameter of 38 mm) operating close to adiabatic conditions, loaded with commercial CaO- and Cu-based materials in pellet form (particle size of about 3 mm). The fixed-bed contained a Cu/CaO molar ratio of about 1.8, which is the theoretical value to accomplish the reduction/calcination with H_2 under neutrally thermal conditions. A fast and complete reduction of CuO with H_2 was observed even at relatively low initial solid bed temperatures (i.e. 400°C). However, only temperatures in the solid bed higher than 700°C allowed a simultaneous reduction/calcination operation, leaving uncalcined material in those zones at lower temperatures. Alarcón et al. [45] evaluated the effect of the fuel gas composition on the CuO reduction/$CaCO_3$ calcination operation. Different Cu/Ca molar ratios were used for this purpose to maintain neutral conditions in the reduction/calcination front. Mixtures of CO and H_2 showed high reactivity with the CuO-based material, resulting in the complete reduction of CuO to Cu in a sharp reaction front and the total oxidation of the gaseous fuel to CO_2 and H_2O. The Cu-based material was able to catalyse the reverse WGS reaction, favoured by the high temperature and the high CO_2 content in the atmosphere. Moreover, combined Ca-Cu oxides formed because of the multicycle operation at high temperature, which slightly modified the chemical composition of the starting materials. These oxides carbonated in the presence of CO_2, affecting the CO_2 sorption capacity of the solid bed. Recently, Fernández et al. [47] studied the reduction/calcination stage using pure CH_4 as reducing gas. The effect of the initial bed temperature and the inlet gas flow rate was evaluated. CuO reduction was favoured when using initial bed temperatures higher than 800°C, resulting in the complete oxidation of inlet CH_4 and the calcination of a large fraction of $CaCO_3$. A low flow rate (i.e. 3 l_N/min of CH_4) allowed a sufficient residence time of the CH_4 inside the reactor to be almost completely converted to CO_2 and H_2O. Temperature profiles higher than 900°C were measured, and large amounts of CO_2 resulting from CH_4 oxidation and $CaCO_3$ decomposition were observed. The relatively long breakthrough periods demonstrated that the reactivity of the CH_4 with the CuO-based material was significantly lower than that measured with H_2.

Consecutive cycles of the three main reaction stages of the Ca-Cu looping process were made by Díez-Martín et al. [19] in a lab-scale fixed-bed reactor (L = 0.2 m, I.D. = 18 mm) under relevant conditions of this process at a large scale. The solid bed contained the three functional materials required to run the process (i.e. a commercial Ni-based catalyst, a $CaO/Ca_{12}A_{11}4O_{33}$ sorbent and a CuO/Al_2O_3 material). The system was able to operate with space velocities of up to 13.5 kg CH_4 h^{-1} kg Ni^{-1} during the SER stage at 675°C and 10 bar producing a gas with more than 93 vol.% H_2 (on a dry basis). The Cu-based solid exhibited fast reduction and oxidation kinetics, but it did not showed any appreciable reactivity towards CH_4 reforming during SER operation. Total O_2 conversion was observed during the Cu oxidation stage. Slightly higher amounts of CO_2 than those predicted by the $CaO/CaCO_3$ equilibrium were measured in the product gas during oxidation due to the carbon deposited during the breakthrough period of the previous SER step. The results obtained along several cycles were highly reproducible, demonstrating the proper chemical stability of the materials. Only a slightly decrease of the CO_2 sorbent capacity was observed. No mixed phases from the different active materials were detected, indicating the absence of any significant chemical interaction between the different solids loaded in the reactor.

5. Process analysis

5.1 Reactor design and modelling

The Ca-Cu looping process was mainly envisaged to be performed in several adiabatic packed-bed reactors operating in parallel. Fixed-bed reactors do not require solid filtering systems downstream since the formation of fines by attrition is avoided, and they allow the operation to take place in a more compact design at a high pressure. Moreover, H_2 and N_2 can be produced at a suitable pressure to be subsequently used in industrial applications and/or power generation. However, pressurized fixed beds require adequate heat management strategies in order to achieve the complete conversion of the solids and at the same time to avoid the formation of hot spots inside the reactors.

The first conceptual design of the overall Ca-Cu process was presented by Fernández et al. [17] in which a quite simple reactor model assuming narrow reaction fronts was used to describe the dynamic performance of every stage of the process. An ideal plug flow model with negligible axial dispersion was considered. Precise operating conditions for the process (i.e. temperature, pressure, steam-to-carbon (S/C) ratio, etc.) and material properties were defined. More rigorous reactor models were latterly developed to describe more precisely the dynamic profiles obtained during every stage of the Ca-Cu process [48, 49]. These are basically 1D models in which a moderate axial mass and heat dispersion are considered and mass and heat transfer resistances between the gas and solid phases are neglected. The model developed by Fernández et al. [50] integrated the kinetic models for the SMR and CaO carbonation reactions occurring during the SER stage. These simulations demonstrated that operating at around 650°C, between 10 and 15 bar, S/C ratios between 3 and 5 and space velocities up to 3.5 kg/m^2 s, allows CH_4 conversions higher than 80% and a product gas with more than 90 vol.% of H_2 to be achieved. A similar reactor model was used to simulate the Cu oxidation stage [51]. In this work, it was theoretically demonstrated that a recirculation of around 80% of the product gas (to dilute the inlet O_2 content to 3–4 vol.%) restricts the temperature achieved in the oxidation front well below 850°C, thereby limiting the $CaCO_3$ calcination (whenever the operation is carried out at pressures higher than 10 bar). The CuO reduction/$CaCO_3$ calcination stage was also simulated in detail by Alarcon and Fernández [52] who demonstrated that appropriate proportions of CuO-based material (which depend on the composition of the reducing gas) provide the heat required for the direct calcination of the carbonated sorbent.

The three main reaction stages of the Ca-Cu process were also modelled in a more recent work by Martini et al. [49] using a relatively complex dynamic model and a simplified model that assumed narrow reaction and heat exchange fronts. The kinetics for the reactions occurring in all the stages of the Ca-Cu process were included. The maximum and minimum values achieved in both temperature and concentration profiles, as well as the reaction and heat exchange fronts velocities calculated using both models, were compared, showing a reasonably good agreement. The operability windows for each reaction stage were identified through sensitivity analyses of the main operating parameters (i.e. the CaO/Cu content in the bed, the composition of the inlet gases, the temperature and the pressure).

In a subsequent work, Fernández and Abanades [53] proposed a new operation strategy to minimize the number of reactors required, increase the CO_2 capture efficiency and avoid possible side reactions (e.g. CaO hydration) that might damage the mechanical characteristics of the Ca-Cu solids. **Figure 3** shows the reactor scheme proposed by these authors. The dynamic operation of the overall Ca-Cu process was simulated assuming that the initial conditions of each reaction stage were the result of the previous step. The simulations showed that the SER operation at 10 bar with temperatures lower than 730°C minimized CaO hydration and the emissions of CO_2. In the reduction/calcination stage, the feed of the fuel (mainly composed of the PSA off-gas resulting from the H_2 purification step downstream of the SER reactor) through the part of the bed that was at the highest temperature led to the complete conversion of the reducing gases. Only five reactors were found to be sufficient to operate the process (L = 6 m, I.D. = 3 m) and produce 30,000 Nm^3/h of H_2, assuming a minimum length/diameter ratio of 2 and a maximum pressure drop of about 10% per stage, which are geometrical constraints and operational limits typically applied to CLC fixed beds [54].

5.2 Process assessment of large-scale Ca-Cu-based plants

The application of the Ca-Cu process having received more attention, due to its good performance in terms of efficiency and CO_2 emissions, is the production of high-purity hydrogen with inherent CO_2 capture. However, its application as CO2 capture process in power plants has been also studied, both applied to the flue gas from a coal-fired power plant [55] and as a pre-combustion CO_2 capture process in a natural gas combined cycle (NGCC) power plant [56]. When applied to a coal-fired power plant, the Ca-Cu process led to a higher electric efficiency than the

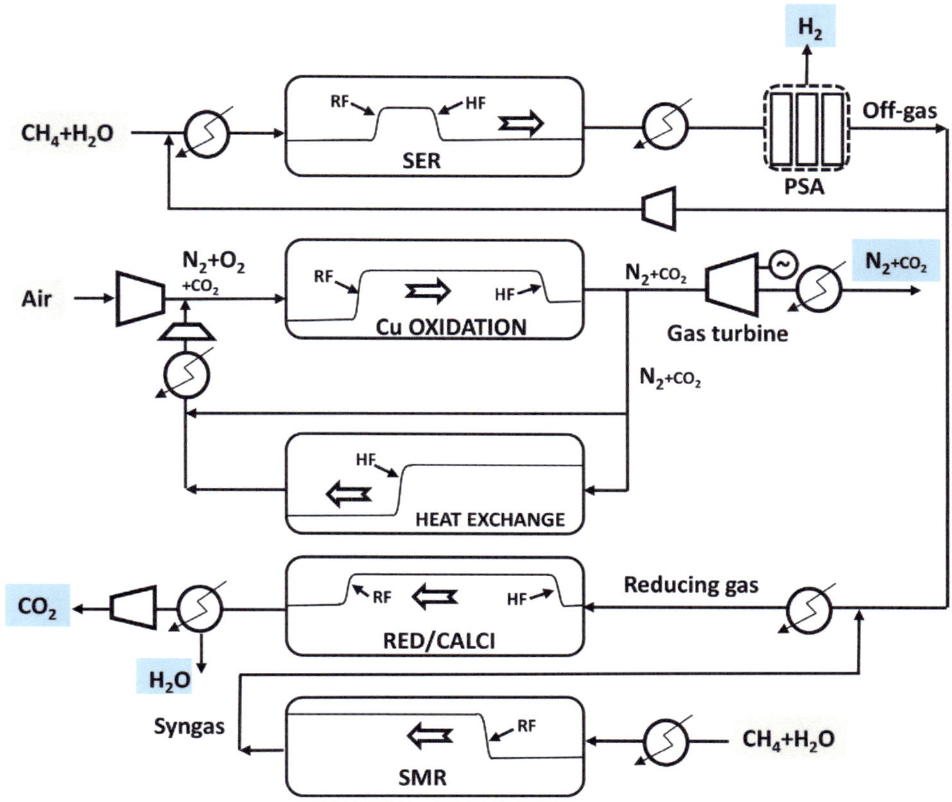

Figure 3.
Scheme of the Ca-Cu looping process for H_2 production (RF and HF refer to reaction front and heat exchange front, respectively).

conventional Ca-Looping and alternative CO_2 capture processes like amine absorption or oxy-combustion. However, despite of this better performance, this scheme relies on a three interconnected fluidized bed reactors system, whose real operation has not been demonstrated. Moreover, the use of pure CH_4 as fuel in the reduction/calcination stage operating at atmospheric pressure is proposed, which penalizes the electric efficiency of the process when compared to the state-of-the-art technology for power production from natural gas (i.e. NGCC). When used as a pre-combustion CO_2 capture process in a NGCC, the Ca-Cu process is operated in the fixed-bed reactors system to produce a high pressure H_2-rich gas in the SER stage that is used as fuel in the gas turbine. In this case, the CO_2 capture efficiency of the Ca-Cu process is totally influenced by the CO_2 capture efficiency of the SER stage, which is limited to around 82% due to the formation of a heat plateau at high temperature within the reactor in the SER stage that limits the carbonation of CaO [57]. In order to improve the CO_2 capture efficiency of the SER stage, Martini et al. [57] evaluated the CO_2 capture efficiency reached through different schemes of the Ca-Cu process and concluded that splitting the SER stage into two steps resulted in the best performance. In this way, the carbon slipped out of the main SER stage is separated in this second step. CO_2 capture efficiency is boosted up to almost 90% using this configuration, which is similar to the benchmark NGCC power plant based on auto-thermal reformer and MDEA absorption process for CO_2 capture (i.e. around 91%). Moreover, electric efficiency of a NGCC power plant with CO_2 capture based on this Ca-Cu scheme has demonstrated to be slightly higher than the electric efficiency of the referred benchmark, which will contribute to a lower electricity cost for this Ca-Cu based NGCC plant.

When focused on large-scale hydrogen production, the performance improvements of the Ca-Cu technology with respect to the commercially ready SMR technology are not as tight as when focused on power production. Martínez et al. [58] evaluated for the first time the performance of a large-scale hydrogen production plant with CO_2 capture using the Ca-Cu process. The simple reactor model based on sharp reaction and heat exchange fronts described in [17] was used for solving the Ca-Cu reactors in this work, which completed the Ca-Cu scheme with the intermediate stages of rinsing, pressurization and depressurization that are needed in a large-scale process. Moreover, the presence of higher hydrocarbons and sulphur compounds in the natural gas used as feedstock made it necessary to include prereforming and desulphurisation stages in the model layout. A total number of 15 reactors was estimated in this work as those needed for running completely a Ca-Cu cycle (i.e. SER-rinse-oxidation-cooling-depressurization-rinse-reduction/calcination-pressurization), having three reactors operating in SER stage, three in the oxidation stage and three in the cooling stage before reduction/calcination, and keeping one reactor for each of the remaining stages. Hydrogen production efficiencies as high as 79% were calculated for the Ca-Cu-based hydrogen production plant in this work, which were reduced to 76% when including the penalties associated to the electricity consumption as well as the benefits for the steam exported.

A more compact reactor design for the Ca-Cu process for hydrogen production was proposed in a later work by Fernández and Abandes [53] who evaluated new operating conditions with the aim of reducing the number of reactors needed. SER stage was operated at a lower pressure (i.e. 11 bar) with an inlet S/C ratio of 3. It was proposed a configuration of only five reactors (i.e. one reactor per each of the Ca-Cu stages, SER, oxidation, cooling, reduction/calcination, cooling/reforming), whose length/diameter ratio was 2 (with a length of 6 m) and the maximum pressure drop allowed was 10% of inlet pressure. The hydrogen efficiency remained unvariable with respect to the value previously reported in [58]. Finally, these performance numbers were completed with an economic analysis by Riva et al. [59]. A rigorous model was used for calculating the fixed-bed reactor system, and it was carried out an optimisation of the pressure drop across the main heat exchangers needed in the plant, as well as across the fixed-bed reactors, with the aim of reducing the H_2 production cost. An economic analysis for a hydrogen production plant based on the Ca-Cu process was carried out for the first time in this work. Each of the four main reactors in the Ca-Cu process (i.e. SER, oxidation, cooling and reduction/calcination) is divided into four sub-reactors for reducing the pressure drop along the reactor and the total amount of functional materials needed to fill the reactors. A sensitivity analysis was performed on the operating pressure of SER and oxidation stages in this work, demonstrating that reducing the operating pressure to 11 bar makes the hydrogen efficiency increase up to 78% and to 79% (i.e. from 74 to 76% when operating at 25 bar) when accounting for electricity and steam exchanges with the surroundings. Considering a common calculation basis of H_2 production of 30,000 Nm^3/h, the calculated cost of hydrogen for the Ca-Cu process ranges between 0.178 and 0.181 €/Nm^3 (operating at 25 and 11 bar, respectively) which is below the cost of 0.194 €/Nm^3 calculated for a benchmark hydrogen production plant based on the well-established SMR technology including CO_2 capture using a MDEA process [59].

One of the inherent advantages of the Ca-Cu concept is the possibility of producing almost pure streams of H_2 and N_2 as part of its products in the SER and oxidation stages, respectively. Such advantage makes it the perfect candidate to be integrated as part of an ammonia production process as recently proposed by Martínez et al. [60]. The schematic of an ammonia production plant based on the Ca-Cu process is shown in **Figure 4**(left). The synthesis gas production island used in the well-established ammonia production process (i.e. consisting of (1) two reforming steps, (2) two

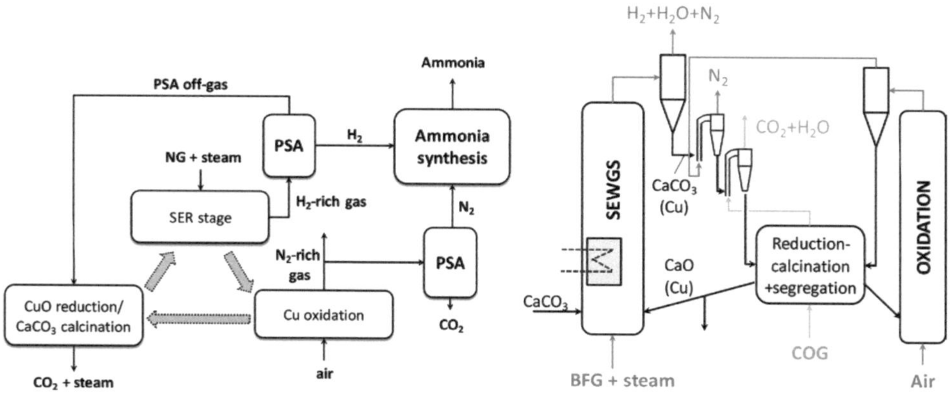

Figure 4.
(Left) Schematic of the Ca-Cu process integrated into an ammonia production plant and (right) simplified Ca-Cu scheme for decarbonising off-gases in a steel mill.

WGS reactors, (3) CO_2 removal section and (4) methanation) is replaced by a fixed-bed Ca-Cu process providing the H_2 and N_2 streams in the right proportion (i.e. 3:1) to be introduced into the NH_3 production loop. Two purification steps would be needed in the Ca-Cu process to remove the impurities of the H_2-rich gas from the SER stage (i.e. unconverted CH_4, CO and CO_2) and from the N_2-rich gas from the non-recirculated gas from the oxidation stage (i.e. CO_2). Based on the analysis done in [60], the ammonia production process integrated with the Ca-Cu process allows reducing the primary energy consumption of a commercial ammonia production plant by around 14%, resulting in 24 GJ/ton_{NH3}. Accounting for the electricity import needed, the advantage of the Ca-Cu-based ammonia plant is maintained. Further research is needed to evaluate the improvements in the ammonia synthesis loop, derived from a higher purity of the H_2/N_2 stream coming from the Ca-Cu process, as well as in the final ammonia production cost.

The potential of the Ca-Cu looping process as a decarbonizing process in the steelmaking sector has been also assessed. Fernández et al. [61] proposed scheme shown in **Figure 4**(right) for decarbonizing a substantial fraction of the blast furnace gas (BFG) produced in the steel mill. A three interconnected fluidized bed system configuration was proposed for this application. In this case, WGS and carbonation of CaO reactions (Eqs. (2) and (3)) occur in the H_2 production stage (i.e. SEWGS) since BFG is mainly composed of CO and CO_2 diluted in N_2, whereas coke oven gas (COG) is used as fuel in the reduction/calcination stage. The circulation of hot solids from an oxidation stage operating at 900°C supplies part of the energy needed for the $CaCO_3$ calcination. A segregation step is needed after the reduction/calcination stage to separate the pure CaO stream demanded by the steelmaking processes and to avoid too much Cu-based solids going to the SEWGS. Using exclusively COG as a fuel for the reduction/calcination, only 30% of the BFG produced in the blast furnace could be decarbonized, whereas adding NG to this reduction/calcination step allows attaining above 91% of CO_2 capture efficiency on the whole steel mill [62].

6. Concluding remarks

The Ca-Cu looping process is an emerging CO_2 capture process that points out an improved efficiency and reduced cost for H_2 production compared to well-established technologies. The functional materials needed for running the process are not a barrier for the progress of the technology, being already available with the proper

Cu and Ca contents. Further research in long-term stability of these materials would be needed in order to elucidate any problem related with activity loss or agglomeration problems. The key reaction stages of the process have been already tested at a sufficient scale under conditions relevant for the operation at a large scale, but investigation in multiple reactor system at a larger scale is needed for corroborating the promising results found at lab-scale.

Author details

Isabel Martínez[1*], Jose R. Fernández[2] and Gemma Grasa[1]

1 Spanish Research Council, ICB-CSIC, Zaragoza, Spain

2 Spanish Research Council, INCAR-CSIC, Oviedo, Spain

*Address all correspondence to: imartinez@icb.csic.es

References

[1] Edenhofer O, Pichs-Madruga R, Sokona Y, Farahani E, Kadner S, Seyboth K, et al. IPCC, 2014: Climate Change 2014: Mitigation of Climate Change. Contribution of Working Group III to the Fifth Assessment Report of the Intergovernmental Panel on Climate Change. Cambridge, United Kingdom and New York, NY, USA: Cambridge University Press; 2014

[2] Olivier JGJ, Janssens-Maenhout G, Muntean M, Peters JAHW. Trends in Global CO_2 Emissions: 2016 Report. The Hague: PBL Netherlands Environmental Assessment Agency. Ispra: European Commission Joint Research Centre; 2016

[3] IEAGHG. Techno-economic evaluation of SMR based standalone (merchant) plant with CCS, 2017/02. 2017

[4] IEA. Technology roadmap. Hydrogen and Fuel Cells. 2015

[5] Boot-Handford ME, Abanades JC, Anthony EJ, Blunt MJ, Brandani S, MacDowell N, et al. Carbon capture and storage update. Energy & Environmental Science. 2014;**7**:130-189. DOI: 10.1039/C3EE42350F

[6] Harrison DP. Sorption-enhanced hydrogen production: A review. Industrial and Engineering Chemistry Research. 2008;**47**:6486-6501. DOI: 10.1021/ie800298z

[7] Di Giuliano A, Gallucci K. Sorption enhanced steam methane reforming based on nickel and calcium looping: A review. Chemical Engineering and Processing: Process Intensification. 2018;**130**:240-252. DOI: 10.1016/J.CEP.2018.06.021

[8] Romano MC, Cassotti EN, Chiesa P, Meyer J, Mastin J. Application of the sorption enhanced-steam reforming process in combined cycle-based power plants. Energy Procedia. 2011;**4**:1125-1132. DOI: 10.1016/j.egypro.2011.01.164

[9] Martínez I, Romano MC, Chiesa P, Grasa G, Murillo R. Hydrogen production through sorption enhanced steam reforming of natural gas: Thermodynamic plant assessment. International Journal of Hydrogen Energy. 2013;**38**:15180-15199. DOI: 10.1016/j.ijhydene.2013.09.062

[10] Martínez I, Murillo R, Grasa G, Rodríguez N, Abanades JC. Conceptual design of a three fluidised beds combustion system capturing CO_2 with CaO. International Journal of Greenhouse Gas Control. 2011;**5**:498-504. DOI: 10.1016/j.ijggc.2010.04.017

[11] Meyer J, Mastin J, Bjørnebøle TK, Ryberg T, Eldrup N. Techno-economical study of the zero emission gas power concept. Energy Procedia. 2011;**4**:1949-1956. DOI: 10.1016/j.egypro.2011.02.075

[12] Reitz M, Junk M, Ströhle J, Epple B. CO_2 Capture in a 300 kWth Indirectly Heated Fluidized Bed Pilot Plant: Operating Experience and Results, 6th High Temperature Solid Looping Cycles Network Meeting, 1-2 September, Milan (Italy); 2015

[13] Abanades JC, Murillo R. Method of capturing CO_2 by means of CaO and the exothermic reduction of a solid. US Patent Number US8506915 B2. 2009

[14] Abanades JC, Murillo R, Fernández JR, Grasa G, Martínez I. New CO_2 capture process for hydrogen production combining Ca and Cu chemical loops. Environmental Science & Technology. 2010;**44**:6901-6904. DOI: 10.1021/es101707t

[15] Fernández JR, Abanades JC. Overview of the Ca-Cu looping process for hydrogen production and/or power generation. Current Opinion

in Chemical Engineering. 2017;**17**:1-8. DOI: 10.1016/j.coche.2017.04.010

[16] Adánez J, Abad A, Garcia-Labiano F, Gayan P, De Diego LF. Progress in chemical-looping combustion and reforming technologies. Progress in Energy and Combustion Science. 2012;**38**:215-282. DOI: 10.1016/j.pecs.2011.09.001

[17] Fernández JR, Abanades JC, Murillo R, Grasa G. Conceptual design of a hydrogen production process from natural gas with CO_2 capture using a Ca-Cu chemical loop. International Journal of Greenhouse Gas Control. 2012;**6**:126-141. DOI: 10.1016/j.ijggc.2011.11.014

[18] Grasa G, Navarro MV, López JM, Díez-Martín L, Fernández JR, Murillo R. Validation of the H_2 production stage via SER under relevant conditions for the Ca/Cu reforming process practical application. Chemical Engineering Journal. 2017;**324**:266-278. DOI: 10.1016/j.cej.2017.04.134

[19] Díez-Martín L, López JM, Fernández JR, Martínez I, Grasa G, Murillo R. Complete Ca/Cu cycle for H_2 production via CH_4 sorption enhanced reforming in a lab-scale fixed bed reactor. Chemical Engineering Journal. 2018;**350**:1010-1021. DOI: 10.1016/J.CEJ.2018.06.049

[20] Liu W, An H, Qin C, Yin J, Wang G, Feng B, et al. Performance enhancement of calcium oxide sorbents for cyclic CO_2 capture—A review. Energy & Fuels. 2012;**26**:2751-2767. DOI: 10.1021/ef300220x

[21] Erans M, Manovic V, Anthony EJ. Calcium looping sorbents for CO_2 capture. Applied Energy. 2016;**180**:722-742. DOI: 10.1016/J.APENERGY.2016.07.074

[22] Shokrollahi Yancheshmeh M, Radfarnia HR, Iliuta MC. High temperature CO_2 sorbents and their application for hydrogen production by sorption enhanced steam reforming process. Chemical Engineering Journal. 2016;**283**:420-444. DOI: 10.1016/j.cej.2015.06.060

[23] Liu W, Dennis JS, Sultan DS, Redfern SAT, Scott SA. An investigation of the kinetics of CO_2 uptake by a synthetic calcium based sorbent. Chemical Engineering Science. 2012;**69**:644-658. DOI: 10.1016/j.ces.2011.11.036

[24] López JM, Grasa G, Murillo R. Evaluation of the effect of inert support on the carbonation reaction of synthetic CaO-based CO_2 sorbents. Chemical Engineering Journal. 2018;**350**:559-572. DOI: 10.1016/J.CEJ.2018.05.014

[25] Arias B, Abanades JC, Grasa GS. An analysis of the effect of carbonation conditions on CaO deactivation curves. Chemical Engineering Journal. 2011;**167**:255-261. DOI: 10.1016/j.cej.2010.12.052

[26] Manovic V, Charland J-P, Blamey J, Fennell PS, Lu DY, Anthony EJ. Influence of calcination conditions on carrying capacity of CaO-based sorbent in CO_2 looping cycles. Fuel. 2009;**88**:1893-1900. DOI: 10.1016/j.fuel.2009.04.012

[27] Donat F, Müller CR. A critical assessment of the testing conditions of CaO-based CO_2 sorbents. Chemical Engineering Journal. 2018;**336**:544-549. DOI: 10.1016/J.CEJ.2017.12.050

[28] Pacciani R, Müller CR, Davidson JF, Dennis JS, Hayhurst AN. Synthetic Ca-based solid sorbents suitable for capturing CO_2 in a fluidized bed. Canadian Journal of Chemical Engineering. 2008;**86**:356-366. DOI: 10.1002/cjce.20060

[29] Koirala R, Reddy GK, Smirniotis PG. Single nozzle flame-made highly

durable metal doped Ca-based sorbents for CO_2 capture at high temperature. Energy and Fuels. 2012;**26**:3103-3109. DOI: 10.1021/ef3004015

[30] Radfarnia HR, Sayari A. A highly efficient CaO-based CO_2 sorbent prepared by a citrate-assisted sol-gel technique. Chemical Engineering Journal. 2015;**262**:913-920. DOI: 10.1016/j.cej.2014.09.074

[31] Kazi SS, Aranda A, Meyer J, Mastin J. High performance CaO-based sorbents for pre- and post- combustion CO_2 capture at high temperature. Energy Procedia. 2014;**63**:2207-2215. DOI: 10.1016/J.EGYPRO.2014.11.240

[32] ASCENT Project. n.d. Available from: http://ascentproject.eu [Accessed: 13 June 2018]

[33] Gayán P, Adánez-Rubio I, Abad A, de Diego LF, García-Labiano F, Adánez J. Development of Cu-based oxygen carriers for chemical-looping with oxygen uncoupling (CLOU) process. Fuel. 2012;**96**:226-238. DOI: 10.1016/J.FUEL.2012.01.021

[34] Imtiaz Q, Broda M, Müller CR. Structure–property relationship of co-precipitated Cu-rich, Al_2O_3- or $MgAl_2O_4$-stabilized oxygen carriers for chemical looping with oxygen uncoupling (CLOU). Applied Energy. 2014;**119**:557-565. DOI: 10.1016/J.APENERGY.2014.01.007

[35] Díez-Martín L, Grasa G, Murillo R, Scullard A, Williams G. Development of suitable CuO-based materials supported on Al_2O_3, $MgAl_2O_4$, and ZrO_2 for Ca/Cu H_2 production process. Industrial and Engineering Chemistry Research. 2018;**57**:2890-2904. DOI: 10.1021/acs.iecr.7b05103

[36] Manovic V, Anthony EJ. CaO-based pellets with oxygen carriers and catalysts. Energy and Fuels. 2011;**25**:4846-4853. DOI: 10.1021/ef2009748

[37] Ridha FN, Lu D, Macchi A, Hughes RW. Combined calcium looping and chemical looping combustion cycles with CaO-CuO pellets in a fixed bed reactor. Fuel. 2015;**153**:202-209. DOI: 10.1016/j.fuel.2015.02.069

[38] Kierzkowska AM, Müller CR. Development of calcium-based, copper-functionalised CO_2 sorbents to integrate chemical looping combustion into calcium looping. Energy & Environmental Science. 2012;**5**:6061-6065. DOI: 10.1039/c2ee03079a

[39] Qin C, Yin J, Liu W, An H, Feng B. Behavior of CaO/CuO based composite in a combined calcium and copper chemical looping process. Industrial and Engineering Chemistry Research. 2012;**51**:12274-12281. DOI: 10.1021/ie300677s

[40] Manovic V, Anthony EJ. Integration of calcium and chemical looping combustion using composite CaO/CuO-based materials. Environmental Science & Technology. 2011;**45**:10750-10756. DOI: 10.1021/es202292c

[41] Manovic V, Wu Y, He I, Anthony EJ. Core-in-shell CaO/CuO-based composite for CO_2 capture. Industrial and Engineering Chemistry Research. 2011;**50**:12384-12391. DOI: 10.1021/ie201427g

[42] Qin C, Yin J, Luo C, An H, Liu W, Feng B. Enhancing the performance of CaO/CuO based composite for CO_2 capture in a combined Ca-Cu chemical looping process. Chemical Engineering Journal. 2013;**28**:75-86. DOI: 10.1016/j.cej.2013.04.115

[43] Kierzkowska AM, Müller CR. Sol-gel-derived, calcium-based, copper-functionalised CO_2 sorbents for an integrated chemical looping combustion-calcium looping CO_2

capture process. ChemPlusChem. 2013;**78**:92-100. DOI: 10.1002/cplu.201200232

[44] Kazi SS, Aranda A, Di Felice L, Meyer J, Murillo R, Grasa G. Development of cost effective and high performance composite for CO_2 capture in Ca-Cu looping process. Energy Procedia. 2017;**114**:211-219. DOI: 10.1016/j.egypro.2017.03.1163

[45] Alarcón JM, Fernández JR, Abanades JC. Study of a Cu-CuO chemical loop for the calcination of $CaCO_3$ in a fixed bed reactor. Chemical Engineering Journal. 2017;**325**:208-220. DOI: 10.1016/j.cej.2017.05.070

[46] Fernández JR, Alarcón JM, Abanades JC. Investigation of a fixed-bed reactor for the calcination of $CaCO_3$ by the simultaneous reduction of CuO with a fuel gas. Industrial and Engineering Chemistry Research. 2016;**55**:5128-5132. DOI: 10.1021/acs.iecr.5b04073

[47] Fernández JR, Alarcón JM, Abanades JC. Study of the calcination of $CaCO_3$ by means of a Cu/CuO chemical loop using methane as fuel gas. Catalysis Today. 2018. (in press). DOI: 10.1016/J.CATTOD.2018.02.053

[48] Fernández JR, Abanades JC, Murillo R. Modeling of sorption enhanced steam methane reforming in an adiabatic fixed bed reactor. Chemical Engineering Science. 2012;**84**:1-11. DOI: 10.1016/j.ces.2012.07.039

[49] Martini M, van den Berg A, Gallucci F, van Sint Annaland M. Investigation of the process operability windows for Ca-Cu looping for hydrogen production with CO_2 capture. Chemical Engineering Journal. 2016;**303**:73-88. DOI: 10.1016/j.cej.2016.05.135

[50] Fernández JR, Abanades JC, Grasa G. Modeling of sorption enhanced steam methane reforming—Part II: Simulation within a novel Ca/Cu chemical loop process for hydrogen production. Chemical Engineering Science. 2012;**84**:12-20. DOI: 10.1016/j.ces.2012.07.050

[51] Fernández JR, Abanades JC, Murillo R. Modeling of Cu oxidation in an adiabatic fixed-bed reactor with N2 recycling. Applied Energy 2014;**113**:1945-1951. DOI:10.1016/j.apenergy.2013.02.013

[52] Alarcón JM, Fernández JR. $CaCO_3$ calcination by the simultaneous reduction of CuO in a Ca/Cu chemical looping process. Chemical Engineering Science. 2015;**137**:254-267. DOI: 10.1016/j.ces.2015.06.030

[53] Fernández JR, Abanades JC. Optimized design and operation strategy of a CaCu chemical looping process for hydrogen production. Chemical Engineering Science. 2017;**166**:144-160. DOI: https://doi.org/10.1016/j.ces.2017.03.039

[54] Spallina V, Chiesa P, Martelli E, Gallucci F, Romano MC, Lozza G, et al. Reactor design and operation strategies for a large-scale packed-bed CLC power plant with coal syngas. International Journal of Greenhouse Gas Control. 2015;**36**:34-50. DOI: 10.1016/j.ijggc.2015.01.025

[55] Ozcan DC, Macchi A, Lu DY, Kierzkowska AM, Ahn H, Müller CR, et al. Ca-Cu looping process for CO_2 capture from a power plant and its comparison with Ca-looping, oxy-combustion and amine-based CO_2 capture processes. International Journal of Greenhouse Gas Control. 2015;**43**:198-212. DOI: 10.1016/J.IJGGC.2015.10.021

[56] Martínez I, Murillo R, Grasa G, Fernández JR, Abanades JC. Integrated combined cycle from natural gas with CO_2 capture using a Ca-Cu chemical loop. AICHE Journal. 2013;**59**:2780-2794. DOI: 10.1002/aic.14054

[57] Martini M, Martínez I, Romano MC, Chiesa P, Gallucci F, van Sint Annaland M. Increasing the carbon capture efficiency of the Ca/Cu looping process for power production with advanced process schemes. Chemical Engineering Journal. 2017;**328**:304-319. DOI: 10.1016/j.cej.2017.07.048

[58] Martínez I, Romano MC, Fernández JR, Chiesa P, Murillo R, Abanades JC. Process design of a hydrogen production plant from natural gas with CO_2 capture based on a novel Ca/Cu chemical loop. Applied Energy. 2014;**114**:192-208. DOI: 10.1016/j.apenergy.2013.09.026

[59] Riva L, Martínez I, Martini M, Gallucci F, van Sint Annaland M, Romano MC. Techno-economic analysis of the Ca-Cu process integrated in hydrogen plants with CO_2 capture. International Journal of Hydrogen Energy. 2018;**43**(33):15720-15738. DOI: 10.1016/J.IJHYDENE.2018.07.002

[60] Martínez I, Armaroli D, Gazzani M, Romano MC. Integration of the Ca-Cu process in Ammonia production plants. Industrial and Engineering Chemistry Research. 2017;**56**:2526-2539. DOI: 10.1021/acs.iecr.6b04615

[61] Fernández JR, Martínez I, Abanades JC, Romano MC. Conceptual design of a Ca-Cu chemical looping process for hydrogen production in integrated steelworks. International Journal of Hydrogen Energy. 2017;**42**:11023-11037. DOI: 10.1016/j.ijhydene.2017.02.141

[62] Martínez I, Fernández JR, Abanades JC, Romano MC. Integration of fluidised bed Ca-Cu chemical looping process in a Steel mill. Energy. 2018;**163**:570-584. DOI:10.1016/j.energy.2018.08.123